T0324713

LINEAR OPERATOR EQUATIONS

Approximation and Regularization

LINEAR OPERATOR EQUATIONS

Approximation and Regularization

M. THAMBAN NAIR

Indian Institute of Technology Madras
India

World Scientific

NEW JERSEY · LONDON · SINGAPORE · BEIJING · SHANGHAI · HONG KONG · TAIPEI · CHENNAI

Published by

World Scientific Publishing Co. Pte. Ltd.

5 Toh Tuck Link, Singapore 596224

USA office: 27 Warren Street, Suite 401-402, Hackensack, NJ 07601

UK office: 57 Shelton Street, Covent Garden, London WC2H 9HE

Library of Congress Cataloging-in-Publication Data
Nair, M. Thamban.
  Linear operator equations : approximation and regularization / by M. Thamban Nair.
    p. cm.
  Includes bibliographical references and index.
  ISBN-13: 978-981-283-564-2 (hard cover : alk. paper)
  ISBN-10: 981-283-564-4 (hard cover : alk. paper)
  1. Linear operators. 2. Operator equations. I. Title.
  QA329.2.N35 2009
  515'.7246--dc22

                                                    2009007531

British Library Cataloguing-in-Publication Data
A catalogue record for this book is available from the British Library.

Copyright © 2009 by World Scientific Publishing Co. Pte. Ltd.

All rights reserved. This book, or parts thereof, may not be reproduced in any form or by any means, electronic or mechanical, including photocopying, recording or any information storage and retrieval system now known or to be invented, without written permission from the Publisher.

For photocopying of material in this volume, please pay a copying fee through the Copyright Clearance Center, Inc., 222 Rosewood Drive, Danvers, MA 01923, USA. In this case permission to photocopy is not required from the publisher.

Printed in Singapore.

Dedicated to

My Mother

Meloth Parvathi Amma

Preface

Many problems in science and engineering have their mathematical formulation as an operator equation

$$Tx = y,$$

where T is a linear or nonlinear operator between certain function spaces. In practice, such equations are solved approximately using numerical methods, as their exact solution may not be often possible or may not be worth looking for due to physical constraints. In such situation, it is desirable to know how the so-called *approximate solution* approximates the *exact solution*, and what would be the error involved in such procedures.

This book is concerned with the investigation of the above theoretical issues related to approximately solving linear operator equations. The main tools used for this purpose are the basic results from functional analysis and some rudimentary ideas from numerical analysis. However, no in-depth knowledge on these disciplines is assumed for reading this book.

Although there are many monographs and survey articles on particular topics under discussion, there exists hardly any book which can be used for a mathematical curriculum to deal with operator theoretic aspects of both *well-posed* and *ill-posed* equations. This book is an attempt to fill this gap so as to be used for a second level course for an M.Sc. programme or for a pre-Ph.D. programme. Such a course will enable students to know how the *important theorems* of functional analysis are in use for solving practical problems that occur in other disciplines.

In the first chapter the concepts of well-posedness and ill-posedness are formally introduced and a few examples of such problems are listed. The second chapter equips the reader with the basics of functional analytic results which have been used throughout the text. The problem of approximately solving *well-posed equations*, in particular, the *second-kind equations*, is treated in the third chapter. The fourth chapter is concerned with the problems associated with *ill-posed equations* and their regularization. In the fifth chapter, ill-posed equations with approximately specified operators have been treated.

Although the book discusses some of the results published in a very recent past, especially, on topics in ill-posed problems, this book is meant only as an introductory text dealing with the issues concerning well-posedness and ill-posedness of linear operator equations. Hence, the topics covered and the discussions carried out in this text are far from exhaustive. Readers interested in the topics under discussion are advised to look into some of the excellent books on the subjects. For instance, the book [5] by Atkinson is a very good reference for second kind operator equations, whereas the books by Baumeister [8], Louis [39], Engl, Hanke and Neubauer [17] and Kirsch [34] give fairly good account on ill-posed operator equations. In fact, there are many journals exclusively devoted to the subject on operator equations, for example, *Integral Equations and Operator Theory, Journal Integral of Equations, Numerical Functional Analysis and Optimization, Inverse Problems, Journal of Inverse and Ill-Posed Problems*, etc., and the readers are encouraged to refer to these journals to have up-to-date status on the topics.

Examples, illustrations, remarks and exercises are interspersed throughout the text. They, along with the problems at the end of each chapter, are intended to help the reader in understanding the concepts under discussion.

Now, some words about numberings and certain notations. Lemmas, propositions, theorems, corollaries, examples and remarks are numbered consecutively within each category using two digits. To mark the end of a proof of a Lemma, Proposition, Theorem, or Corollary, we use the symbol □ while for marking the end of a Remark and Example, the symbol ◊ is used. **Bold face** is used when a new terminology is defined, and *italics* are used to emphasize a terminology or a statement.

Acknowledgments: This book had its origin in the form of *Notes on Linear Operator Equations* prepared for my personal use during my second visit to the Centre for Mathematics and its Applications (CMA), Australian National University, Canberra, during June–December 1993. My interaction and collaboration with Professors Bob Anderssen and Markus Hegland during that visit were of immense input for sustaining my interest in the area of ill-posed problems. Subsequently, the notes grew in size along with my own contributions to the field and my collaborations with my doctoral students and some of the experts in the area of numerical functional analysis, Professors Eberhard Schock (University of Kaiserslautern, Germany), Ulrich Tautenhahn (University of Zittau, Germany), and Sergei Pereverzev (Johan Radon Institute for Applied and Computational Mathematics, Linz, Austria). The encouragement I received from Professor Schock during my visits to his university many times during the last 16 years helped me a lot in laying the ground work to bring out my notes in the form of a book.

I gratefully acknowledge the support received from my institution, IIT Madras, under the *Golden Jubilee Book Writing Scheme* during the finalization of this book, and *World Scientific* for its publication. My thanks are also due to Dr. P.S. Srinivasan of Bharathidasan University for helping me to set properly the style-files of the publisher.

Some of my doctoral students read parts of my *Notes* and pointed out many corrections. I am thankful to all of them, especially Pallavi Mahale, Ravishankar, Deepesh and Zeeta Paul.

I thank my wife Sunita and daughters Priya and Sneha for their constant support and for having to bear my absence for too many hours, especially in many of the evenings and holidays in the last few years, while this book was in preparation.

I am always thankful to my mother Meloth Parvathi Amma for her continuous blessings throughout my life, to whom this book is dedicated.

M. Thamban Nair

Contents

Chapter 1

Introduction

1.1 General Introduction

Abstract model for many problems in science and engineering take the form of an operator equation

$$Tx = y, \tag{1.1}$$

where $T : X \to Y$ is a linear or nonlinear operator (between certain function spaces or Euclidean spaces) such as a differential operator or integral operator or a matrix. The spaces X and Y are linear spaces endowed with certain *norms* on them. The first and foremost question that one raises about the operator equation (1.1) is whether a solution exists in X for a given $y \in Y$. Once this question is answered affirmatively, then the next question is that of the uniqueness of the solution. Third question, which is very important in view of its application, is whether the solution depends continuously on the data y; that is, if y is perturbed slightly to, say $\tilde{y}$, then, is the corresponding solution $\tilde{x}$ close to x? If all the above three questions are answered positively, then (according to Hadamard [29]) the problem of solving the equation (1.1) is *well-posed*; otherwise it is *ill-posed*.

We plan to study linear equations, that is, equations of the form (1.1) with $T : X \to Y$ being a linear operator between normed linear spaces X and Y, and methods for solving them approximately. In the following chapters, we quote often the case when T is a compact operator or T is of the form $\lambda I - A$, where A is a compact operator and λ is a nonzero scalar. Prototype of a compact operator K that we shall discuss at length is the Fredholm integral operator defined by

$$(Kx)(s) = \int_a^b k(s,t)x(t)dt, \quad s,t \in [a,b],$$

1

where x belongs to either $C[a, b]$, the space of continuous functions on $[a, b]$, or $L^2[a, b]$, the space of all square integrable functions on $[a, b]$ with respect to the Lebesgue measure on $[a, b]$, and $k(\cdot, \cdot)$ is a nondegenerate kernel. Thus, examples of equations of the form (1.1) include Fredholm integral equations of the first kind,

$$\int_\Omega k(s, t)x(t)dt = y(s), \quad s, t \in \Omega, \tag{1.2}$$

and Fredholm integral equations of the second kind,

$$\lambda x(s) - \int_\Omega k(s, t)x(t)dt = y(s), \quad s, t \in \Omega. \tag{1.3}$$

We shall see that equation (1.2) is an ill-posed whereas equation (1.3) is well-posed whenever λ is not an eigenvalue of K.

Now we shall formally define the well-posedness and ill-posedness of equation (1.1).

1.2   Well-Posedness and Ill-Posedness

Let X and Y be linear spaces over the scalar field $\mathbb{K}$ of real or complex numbers, and let $T : X \to Y$ be a linear operator. For $y \in Y$, consider the equation (1.1). Clearly, (1.1) has a solution if and only if $y \in R(T)$, where

$$R(T) := \{Tx : x \in X\}$$

is the *range space* of the operator T. Also, we may observe that (1.1) can have at most one solution if and only if $N(T) = \{0\}$, where

$$N(T) := \{x \in X : Tx = 0\}$$

is the *null space* of the operator T.

If the linear spaces X and Y are endowed with norms $\| \cdot \|_X$ and $\| \cdot \|_Y$ respectively, then we can talk about continuous dependence of the solution; i.e., if $\tilde{y} \in Y$ is such that $\|y - \tilde{y}\|_Y$ is 'small', and if x and $\tilde{x}$ satisfy $Tx = y$ and $T\tilde{x} = \tilde{y}$, respectively, then we can enquire whether $\|x - \tilde{x}\|_X$ is also 'small'.

Throughout this book, norms on linear spaces will be denoted by $\| \cdot \|$ irrespective of which space under consideration, except in certain cases where it is necessary to specify them explicitly.

Let X and Y be normed linear spaces and $T : X \to Y$ be a linear operator.

We say that equation (1.1) or the problem of solving the equation (1.1) is **well-posed** if

(a) for every $y \in Y$, there exists a unique $x \in X$ such that $Tx = y$, and

(b) for every $y \in Y$ and for every $\varepsilon > 0$, there exists $\delta > 0$ with the following properties: If $\tilde{y} \in Y$ with $\|\tilde{y} - y\| \leq \delta$, and if $x, \tilde{x} \in X$ are such that $Tx = y$ and $T\tilde{x} = \tilde{y}$, then $\|x - \tilde{x}\|_X \leq \varepsilon$.

Condition (a) in the above definition is the assertion of existence and uniqueness of a solution of (1.1), and (b) asserts the continuous dependence of the solution on the data y.

If equation (1.1) or the problem of solving (1.1) is not well-posed, then (1.1) is called an **ill-posed** equation or an ill-posed problem.

We may observe that equation (1.1) is well-posed if and only if the operator T is bijective and the inverse operator $T^{-1} : Y \to X$ is continuous. We shall see later that if X and Y are Banach spaces, i.e., if X and Y are complete with respect to the metrics induced by their norms, and if T is a continuous linear operator, then continuity of T^{-1} is a consequence of the fact that T is bijective.

It may be interesting to notice that the condition (a) in definition of well-posedness is equivalent to the following:

(c) There exists $\delta_0 > 0$ and $y_0 \in Y$ such that for every $y \in Y$ with $\|y - y_0\|_Y < \delta_0$, there exists a unique $x \in X$ satisfying $Tx = y$.

The equivalence of (a) and (c) is a consequence of the facts that the range of T is a subspace, and a subspace can contain an open ball if and only if it is the whole space.

Exercise 1.1. Prove the last statement.

Next we mention a few examples of well-posed as well as ill-posed equations which arise in practical situations. Our claims, that a quoted example is in fact well-posed or ill-posed, can be justified after going through some of the basics in functional analysis given in Chapter 2.

1.2.1 *Examples of well-posed equations*

• Love's equation in electrostatics (Love [40]):

$$x(s) - \frac{1}{\pi} \int_{-1}^{1} \frac{x(t)}{1 + (t-s)^2} dt = 1, \quad s \in [-1, 1].$$

• A singular integral equation in the theory of intrinsic viscosity (Polymer physics) (Kirkwood and Riseman [33]):

$$\lambda x(s) - \int_{-1}^{1} \frac{1}{|s-t|^{\frac{1}{2}}} x(t) dt = s^2, \quad 0 < \alpha < 1, \quad s \in [-1,1].$$

- Two-dimensional Dirichlet problem from potential theory (Kress [35]):

$$x(s) - \int_{0}^{2\pi} k(s,t) x(t) dt = -2y(s), \quad s \in [0, 2\pi],$$

where

$$k(s,t) = -\frac{ab}{\pi[a^2 + b^2 - (a^2 - b^2)\cos(s+t)]}, \quad a > 0, b > 0.$$

1.2.2 *Examples of ill-posed equations*

- Geological prospecting (Groetsch [26]):

$$\gamma \int_{0}^{1} \frac{1}{[1 + (s-t)^2]^{3/2}} x(t) dt = y(s), \quad s \in [0,1].$$

- Backward heat conduction problem (Groetsch [26]):

$$\int_{0}^{\pi} k(s,t) x(t) dt = y(s), \quad s \in [0, \pi],$$

where

$$k(s,t) = \frac{2}{\pi} \sum_{n=1}^{\infty} e^{-n^2} \sin(ns) \sin(nt).$$

- Computerized tomography (Abel equation of the first kind) (Engl [15]):

$$\int_{s}^{R} \frac{2tx(t)}{\sqrt{t^2 - s^2}} dt = y(s), \quad s \in (0, R].$$

1.3   What Do We Do in This Book?

For well-posed equations, the situation where the data y is known only approximately, say $\tilde{y}$ in place of y, is taken care. But in examples which arise in applications, the operator T may also be known only approximately, or we may approximate it for the purpose of numerical computations. Thus, what one has is an operator $\tilde{T}$ which is an approximation of T. In such case, it is necessary to ensure the existence of a unique solution $\tilde{x}$ for the equation $\tilde{T}\tilde{x} = \tilde{y}$, and then to guarantee that $x - \tilde{x}$ is 'small' whenever

$T - \tilde{T}$ and $y - \tilde{y}$ are 'small'. This aspect of the problem has been studied in Chapter 3 with special emphasis on the second kind operator equation

$$\lambda x - Ax = y,$$

where A is a bounded operator on a Banach space X and λ is a nonzero scalar which is not in the spectrum of A. In the special case of A being a compact operator, the above requirement on λ is same as that it is not an eigenvalue of A.

If equation (1.1) has no solution, then the next best thing one can think of is to look for a unique element with some prescribed properties which minimizes the residual error $\|Tx - y\|$, and then enquire whether the modified problem is well-posed or not. If it is still ill-posed, then the need of regularization of the problem arises. In a regularization, the original problem is replaced by a family of well-posed problems, depending on certain parameter. Proper choice of the parameters yielding convergence and order optimal error estimates is crucial aspect of the regularization theory. These aspects of the problem have been considered in Chapter 4.

In Chapter 5, we use the approximation methods considered in Chapter 3 to study the ill-posed problems when the operator under consideration is also known only approximately. This chapter includes some of the new results on integral equations of the first kind which have not appeared so far in the literature.

Discussion on all the above considerations require a fare amount of results from Functional Analysis and Operator Theory. The purpose of Chapter 2 is to introduce some of the results in this regard.

Chapter 2

Basic Results from Functional Analysis

In this chapter we recall some of the basic concepts and results from Functional Analysis and Operator Theory. Well-known results are stated without proofs. These concepts and results are available in standard textbooks on Functional Analysis, for example the recent book [51] by the author. However, we do give detailed proofs of some of the results which are particularly interested to us in the due course.

2.1 Spaces and Operators

2.1.1 *Spaces*

Let X be a linear space (or vector space) over $\mathbb{K}$, the field of real or complex numbers. Members of X are called **vectors** and members of $\mathbb{K}$ are called **scalars**.

A linear space endowed with a norm is called a **normed linear space**. Recall that a **norm** on a linear space X is a non-negative real-valued function

$$x \mapsto \|x\|, \quad x \in X,$$

which satisfies the following conditions:

(i) $\forall x \in X, \ \|x\| = 0 \iff x = 0,$
(ii) $\|\alpha x\| = |\alpha| \|x\| \ \ \forall x \in X, \ \ \forall \alpha \in \mathbb{K},$
(iii) $\|x + y\| \leq \|x\| + \|y\| \ \ \forall x, y \in X.$

It is easily seen that the map

$$(x, y) \mapsto \|x - y\|, \quad (x, y) \in X \times X,$$

is a metric on X. It also follows from the inequality (iii) above that

$$\|x - y\| \geq \big| \|x\| - \|y\| \big| \quad \forall x, y \in X,$$

so that the map $x \mapsto \|x\|$, $x \in X$, is a uniformly continuous function from X to $\mathbb{K}$ with respect to the above metric on X.

In due course, convergence of sequences in a normed linear space and continuity of functions between normed linear spaces will be with respect to the above referred metric induced by the norm on the spaces.

NOTATION: For a convergent sequence (s_n) in a subset of a metric space with $\lim\limits_{n \to \infty} s_n = s$, we may simply write '$s_n \to 0$' instead of writing '$s_n \to 0$ as $n \to \infty$'.

If a normed linear space is complete with respect to the induced metric, then it is called a **Banach space**.

Now, we consider a linear space X with an **inner product** $\langle \cdot, \cdot \rangle$ on it, that is, a $\mathbb{K}$-valued map

$$(x, y) \mapsto \langle x, y \rangle$$

on $X \times X$ satisfying

(i) $\langle x, x \rangle \geq 0 \quad \forall\, x \in X,$
(ii) $\forall x \in X, \ \langle x, x \rangle = 0 \iff x = 0,$
(iii) $\langle \alpha x, y \rangle = \alpha \langle x, y \rangle \quad \forall \alpha \in \mathbb{K}, \ \forall x, y \in X,$
(iv) $\langle x + y, u \rangle = \langle x, u \rangle + \langle y, u \rangle \quad \forall x, y, u \in X,$
(v) $\langle x, y \rangle = \overline{\langle y, x \rangle} \quad \forall x, y \in X.$

Here $\overline{\alpha}$ denotes the complex conjugate of α.

A linear space together with an inner product on it is called an **inner product space**. An important inequality on an inner product space X is the *Schwarz inequality*, also known as *Cauchy–Schwarz inequality*.

Schwarz inequality: For every x and y in an inner product space,

$$|\langle x, y \rangle|^2 \leq \langle x, x \rangle \langle y, y \rangle. \tag{2.1}$$

Using the Schwarz inequality, it can be seen that the map

$$x \mapsto \|x\| := \langle x, x \rangle^{1/2}, \quad x \in X,$$

defines a norm on X. If X is complete with respect to the metric induced by this norm, then it is called a **Hilbert space**.

The reader is advised to verify the assertions in the following examples.

Example 2.1. For $x = (\alpha_1, \ldots, \alpha_n)$ in $\mathbb{K}^n$, let

$$\|x\|_1 = \sum_{j=1}^{n} |\alpha_j| \quad \text{and} \quad \|x\|_\infty := \max_{1 \le j \le n} |\alpha_j|.$$

Then, it can be easily seen that $\| \cdot \|_1$ and $\| \cdot \|_\infty$ are norms on $\mathbb{K}^n$, and $\mathbb{K}^n$ is a Banach space with respect to these norms. Also, we see that $\langle \cdot, \cdot \rangle$ defined by

$$\langle x, y \rangle := \sum_{j=1}^{n} \alpha_j \overline{\beta_j}$$

for $x = (\alpha_1, \ldots, \alpha_n)$, $x = (\beta_1, \ldots, \beta_n)$ in $\mathbb{K}^n$ is an inner product on X, which induces the norm

$$\|x\|_2 = \left(\sum_{j=1}^{n} |\alpha_j|^2 \right)^{1/2}, \quad x = (\alpha_1, \ldots, \alpha_n) \in \mathbb{K}^n.$$

With respect to the above inner product, $\mathbb{K}^n$ is a Hilbert space. ◊

Note that, the Schwarz inequality (2.1) on $\mathbb{K}^n$ with respect to inner product in the above example takes the form

$$\sum_{j=1}^{n} |\alpha_j \overline{\beta_j}| \le \left(\sum_{j=1}^{n} |\alpha_j|^2 \right)^{1/2} \left(\sum_{j=1}^{n} |\beta_j|^2 \right)^{1/2}$$

for $x = (\alpha_1, \ldots, \alpha_n)$, $x = (\beta_1, \ldots, \beta_n)$ in $\mathbb{K}^n$, which is a consequence of the inequality $2ab \le a^2 + b^2$ for $a, b \in \mathbb{R}$.

A more general form of such inequality is the *Hölder's inequality*:

Hölder's inequality: For $x = (\alpha_1, \ldots, \alpha_n)$, $x = (\beta_1, \ldots, \beta_n)$ in $\mathbb{K}^n$ and $1 < p < \infty$,

$$\sum_{j=1}^{n} |\alpha_j \overline{\beta_j}| \le \left(\sum_{j=1}^{n} |\alpha_j|^p \right)^{1/p} \left(\sum_{j=1}^{n} |\beta_j|^q \right)^{1/q}, \tag{2.2}$$

where $q > 0$ is such that $p + q = pq$ (cf. [51], section 2.1.1).

Example 2.2. Let $1 < p < \infty$. For $x = (\alpha_1, \ldots, \alpha_n)$ in $\mathbb{K}^n$, let

$$\|x\|_p = \left(\sum_{j=1}^{n} |\alpha_j|^p \right)^{1/p}.$$

Using the Hölder's inequality (2.2), it can be shown that

$$\|x\|_p = \Big(\sum_{j=1}^{n} |\alpha_j|^p \Big)^{1/p}, \quad x = (\alpha_1, \ldots, \alpha_n) \in \mathbb{K}^n$$

also defines a norm on $\mathbb{K}^n$ for $1 < p < \infty$ and $\mathbb{K}^n$ is a Banach space with respect to this norm. For $1 \leq p \leq \infty$, we shall denote the space $\mathbb{K}^n$ with the norm $\|\cdot\|_p$ by $\ell^p(n)$.

More generally, let X be a finite dimensional linear space, and let $E = \{u_1, \ldots, u_n\}$ be a basis of X. For $x \in X$, let $\alpha_1(x), \ldots, \alpha_n(x)$ be the scalars such that

$$x = \sum_{j=1}^{n} \alpha_j(x) u_j.$$

For $1 \leq p \leq \infty$, let

$$\|x\|_{E,p} = \begin{cases} \Big(\sum_{j=1}^{n} |\alpha_j(x)|^p \Big)^{1/p} & \text{if } p < \infty \\ \max_{1 \leq j \leq n} |\alpha_j(x)| & \text{if } p = \infty. \end{cases}$$

Then, as in the case of $\mathbb{K}^n$, $\|\cdot\|_{E,p}$ is a norm on X, and X is a Banach space with respect to this norm. Also, the map

$$(x, y) \mapsto \langle x, y \rangle := \sum_{j=1}^{n} \alpha_j(x) \overline{\alpha_j(y)}, \quad x, y \in X,$$

defines an inner product on X which makes it a Hilbert space. $\Diamond$

Example 2.3. (Space $\ell^p(\mathbb{N})$) Let $1 \leq p \leq \infty$ and for a sequence $x = (\alpha_n)$ of scalars, let

$$\|x\|_p = \begin{cases} \Big(\sum_{j=1}^{\infty} |\alpha_j|^p \Big)^{1/p} & \text{if } p < \infty \\ \max_{j \in \mathbb{N}} |\alpha_j| & \text{if } p = \infty. \end{cases}$$

Let $\ell^p(\mathbb{N})$ or simply ℓ^p, be the set of all sequences $x = (\alpha_n)$ such that $\|x\|_p < \infty$. It can be easily seen that $\|\cdot\|_1$ and $\|\cdot\|_\infty$ are norms on ℓ^1 and ℓ^∞, respectively. To see that $\|\cdot\|_p$ is norm on ℓ^p for $1 < p < \infty$, one may make use of the Hölder's inequality (2.2) to obtain

$$\sum_{j=1}^{\infty} |\alpha_j \overline{\beta_j}| \leq \Big(\sum_{j=1}^{\infty} |\alpha_j|^p \Big)^{1/p} \Big(\sum_{j=1}^{\infty} |\beta_j|^q \Big)^{1/q}$$

for scalar sequences (α_n) and (β_n), where $q > 0$ satisfies $p + q = pq$. It can be shown that ℓ^p is a Banach space with respect to the norm $\| \cdot \|_p$. Also, for $x = (\alpha_n)$ and $y = (\beta_n)$ in ℓ^2, the map

$$(x, y) \mapsto \langle x, y \rangle := \sum_{j=1}^{\infty} \alpha_j \overline{\beta_j}$$

defines an inner product on ℓ^2 which induces the norm $\| \cdot \|_2$ and makes it a Hilbert space. ◇

Example 2.4. (Space $C[a, b]$) Consider the linear space $C[a, b]$ of all $\mathbb{K}$-valued continuous functions defined on $[a, b]$. For $x \in C[a, b]$ and $1 \leq p \leq \infty$, let

$$\|x\|_p = \begin{cases} \left(\displaystyle\int_a^b |x(t)|^p dt \right)^{1/p} & \text{if } p < \infty \\ \sup_{a \leq t \leq b} |x(t)| & \text{if } p = \infty. \end{cases}$$

It can be easily seen that $\| \cdot \|_1$ and $\| \cdot \|_\infty$ are norms on $C[a, b]$. To see that $\| \cdot \|_p$ is also a norm on $C[a, b]$, one may make use of the Hölder's inequality (2.2) on $C[a, b]$, namely,

$$\int_a^b |x(t)y(t)| \, dt \leq \left(\int_a^b |x(t)|^p \, dt \right)^{1/p} \left(\int_a^b |x(t)|^p \, dt \right)^{1/q}$$

for $x, y \in C[a, b]$. It can be seen that the metric induced by $\| \cdot \|_p$ is complete only for $p = \infty$. Also, the map

$$(x, y) \mapsto \langle x, y \rangle := \int_a^b x(t) \overline{y(t)} dt$$

defines an inner product on $C[a, b]$. Since the induced norm $\| \cdot \|_2$ is not complete, $C[a, b]$ with the above inner product is not a Hilbert space. ◇

Example 2.5. (Space $L^p[a, b]$) For $1 \leq p < \infty$, the completion of the space $C[a, b]$ with respect to the norm $\| \cdot \|_p$ is linearly isometric with $L^p[a, b]$, the space of all Lebesgue measurable functions x on $[a, b]$ such that

$$\int_a^b |x(t)|^p dt < \infty.$$

Here, the integration is with respect to Lebesgue measure, and equality of elements in $L^p[a, b]$ is defined by

$$f = g \iff f(t) = g(t) \quad \text{for almost all } t \in [a, b].$$

The norm on $L^p[a, b]$ is defined by

$$\|x\|_p = \Big(\int_a^b |x(t)|^p dt \Big)^{1/p}, \quad x \in L^p[a, b].$$

On the space $L^2[a, b]$,

$$\langle x, y \rangle = \int_a^b x(t)\overline{y(t)}dt, \quad x, y \in L^2[a, b],$$

defines an inner product which makes it a Hilbert space.

More generally, let $(\Omega, \mathcal{A}, \mu)$ be a measure space, that is, $\mathcal{A}$ is a σ-algebra of subsets of a set Ω and μ is a measure on $(\Omega, \mathcal{A})$. Then, defining

$$\|x\|_p = \Big(\int_\Omega |x(t)|^p dt \Big)^{1/p}$$

for measurable functions x on Ω, we have the general form of Hölder's inequality as

$$\int_\Omega |x(t)y(t)| \, d\mu(t) \le \|x\|_p \|y\|_q \tag{2.3}$$

for any two measurable functions x, y on Ω. In this case, it is known that

$$L^p(\Omega, \mathcal{A}, \mu) := \{x : x \text{ measurable on } \Omega, \|x\|_p < \infty\}$$

is a Banach space with respect to the norm $\|\cdot\|_p$ (cf. Rudin [66]).   ◊

We remarked in Example 2.1 that a finite dimensional space with norm $\|\cdot\|_{E,p}$ is a Banach space. One may ask whether this is true with respect to any arbitrary norm. The answer is affirmative as the following theorem shows (cf. [51], Theorem 2.25).

Theorem 2.1. *Every finite dimensional subspace of a normed linear space is complete.*

The above theorem can be proved by first showing that a finite dimensional linear space is complete with respect to a particular norm, say $\|\cdot\|_{E,\infty}$ defined as in Example 2.2, and then use the following result (cf. see [51], Theorem 2.24).

Theorem 2.2. *Any two norms on a finite dimensional linear space are equivalent.*

We may recall that two norms, say $\| \cdot \|$ and $\| \cdot \|_*$ on a linear space X are equivalent, if there exist positive numbers a and b such that

$$a\|x\| \leq \|x\|_* \leq b\|x\| \quad \forall\, x \in X.$$

CONVENTION: Unless otherwise specified, the norm and inner product on any linear space will be denoted simply by $\| \cdot \|$ and $\langle \cdot, \cdot \rangle$ respectively, and the metric under discussion is the one induced by the corresponding norm. On $\mathbb{K}$, we always take the standard norm defined by the absolute value. In the sequel, the closure of a subset S of a normed linear space is denoted by $\mathrm{cl}(S)$ or $\overline{S}$.

The notion of **orthogonality** of vectors is very important in an inner product space. Elements x, y in an inner product space are said to be **orthogonal** if $\langle x, y \rangle = 0$, and in that case we may write $x \perp y$. Associated with orthogonality, we have the following identity:

Pythagoras theorem: Let X be an inner product space and $x, y \in X$. Then

$$\|x + y\|^2 = \|x\|^2 + \|y\|^2 \quad \text{whenever} \quad x \perp y. \tag{2.4}$$

For a subset S of X, we write

$$S^\perp := \{x \in X : \langle x, u \rangle = 0\ \forall\, u \in S\}.$$

It can be seen that $S^\perp$ is a closed subspace of X and $S \cap S^\perp \subseteq \{0\}$. If S_1 and S_2 are subsets of an inner product space such that $x \perp y$ for every $x \in S_1$ and $y \in S_2$, then we write $S_1 \perp S_2$.

A subset E of X is said to be an **orthogonal set** if $x \perp y$ for every distinct $x, y \in E$, and it is said to be an **orthonormal set** if it is orthogonal and $\|x\| = 1$ for every $x \in E$. It can be seen that every orthogonal set which does not contain the zero element is linearly independent. In particular, every orthonormal set is linearly independent. A maximal orthonormal set is called an **orthonormal basis**. It can be seen that an orthonormal set E is an orthonormal basis if and only if $E^\perp = \{0\}$, that is, if and only if, for every $x \in X$, $x \perp u$ for every $u \in E$ implies $x = 0$.

A sequence (x_n) in X is said to be an **orthonormal sequence** if the set $\{x_n : n \in \mathbb{N}\}$ is an orthonormal set.

The following two theorems are well known ([51], Theorem 4.9 and Theorem 4.10).

Theorem 2.3. *Let X be a Hilbert space and (u_n) be an orthonormal sequence in X. Then the following are equivalent.*

(i) $E = \{u_n : n \in \mathbb{N}\}$ *is an orthonormal basis.*

(ii) *For every* $x \in X$, $x = \sum_{n=1}^{\infty} \langle x, u_n \rangle u_n$.

(iii) *For every* $x \in X$, $\|x\|^2 = \sum_{n=1}^{\infty} |\langle x, u_n \rangle|^2$.

Theorem 2.4. *A Hilbert space* X *is separable if and only if every orthonormal basis of* X *is countable.*

The equalities in Theorem 2.3 (ii) and (iii) are called **Fourier expansion** and **Parseval's formula**, respectively.

2.1.2   *Bounded operators*

Recall that a function $T : X \to Y$ between linear spaces X and Y is called a **linear operator** if

$$T(x + y) = T(x) + T(y) \quad \forall\, x, y \in X,$$

$$T(\alpha x) = \alpha T(x) \quad \forall\, x \in X, \alpha \in \mathbb{K}.$$

If $T : X \to Y$ is a linear operator, then we shall often write Tx instead of $T(x)$ for $x \in X$. We note that

$$N(T) = \{x \in X : Tx = 0\}$$

is a subspace of X, called the **null space** of T, and

$$R(T) = \{Tx : x \in X\}$$

is a subspace of Y, called the **range** of T. It is immediate that a linear operator $T : X \to Y$ is one-to-one or injective if and only if $N(T) = \{0\}$, and it is onto or surjective if and only if $R(T) = Y$.

Let $T : X \to Y$ be a linear operator between normed linear spaces X and Y. It can be seen that T is continuous if and only if there exists $c > 0$ such that

$$\|Tx\| \leq c\,\|x\| \quad \forall\, x \in X,$$

and in that case

$$\inf\{c > 0 : \|Tx\| \leq c\,\|x\| \,\forall\, x \in X\} = \sup\{\|Tx\| : x \in X, \|x\| \leq 1\}.$$

Moreover, T is continuous if and only if T maps every bounded subsets of X onto bounded subsets of Y. Therefore, a continuous linear operator is also called a **bounded linear operator** or simply a **bounded operator**.

We denote the set of all bounded operators from X to Y by $\mathcal{B}(X,Y)$. If $Y = X$ then we write $\mathcal{B}(X,Y)$ as $\mathcal{B}(X)$. If $T \in \mathcal{B}(X)$, then we say that T is a **bounded operator on** X.

It is seen that $\mathcal{B}(X,Y)$ is a linear space with addition and scalar multiplication are defined pointwise, that is, if T, T_1, T_2 are in $\mathcal{B}(X,Y)$ and $\alpha \in \mathbb{K}$, then $T_1 + T_2$ and αT are defined by

$$(T_1 + T_2)(x) = T_1 x + T_2 x, \quad x \in X,$$

$$(\alpha T)(x) = \alpha T x, \quad x \in X.$$

Also, the map $T \mapsto \|T\|$ defined by

$$\|T\| := \sup\{\|Tx\| : x \in X, \|x\| \le 1\}, \quad T \in \mathcal{B}(X,Y),$$

is a norm on $\mathcal{B}(X,Y)$. If we want to specify the spaces under consideration we write $\|T\|$ as $\|T\|_{X \to Y}$. We observe that for $T \in \mathcal{B}(X,Y)$,

$$\|Tx\| \le \|T\| \, \|x\|, \quad \forall x \in X,$$

and if $c > 0$ is such that $\|Tx\| \le c\|x\|$ for all $x \in X$, then $\|T\| \le c$. Hence, if there exists $c \ge 0$ and a nonzero vector $x_0 \in X$ such that $\|Tx\| \le c\|x\|$ for all $x \in X$ and $\|Tx_0\| = c\|x_0\|$, then it follows that $\|T\| = c_0$.

If Y is a Banach space, then $\mathcal{B}(X,Y)$ is also a Banach space (cf. [51], Theorem 3.12). If Y is the scalar field $\mathbb{K}$, then the space $\mathcal{B}(X,\mathbb{K})$ is called the **dual** of X and it is denoted by X'. Elements of X' are called **continuous** or **bounded linear functionals** on X. Usually, elements of X' are denoted by smaller case letters f, g, etc.

Let X, Y and Z be normed linear spaces, and $T_1 \in \mathcal{B}(X,Y)$ and $T_2 \in \mathcal{B}(Y,Z)$. Then it follows that

$$T_2 T_1 \in \mathcal{B}(X,Z) \quad \text{and} \quad \|T_2 T_1\| \le \|T_2\| \, \|T_1\|.$$

Here, the operator $T_2 T_1$ is defined as the composition of T_2 and T_1, i.e.,

$$(T_2 T_1)(x) = T_2(T_1 x), \quad x \in X.$$

From the above observations, it follows that if $T \in \mathcal{B}(X,Y)$, then for every $f \in Y'$, the map $g_f : X \to \mathbb{K}$ defined by

$$g_f(x) = f(Tx), \quad x \in X,$$

belongs to X'. Moreover, the map $T' : Y' \to X'$ defined by

$$T'f = g_f, \quad f \in Y',$$

belongs to $\mathcal{B}(Y',X')$. The operator T' is called the **transpose** of T. We see that $\|T'\| \le \|T\|$.

Example 2.6. Let (a_{ij}) be an $m \times n$ matrix of scalars and let $T : \mathbb{K}^n \to \mathbb{K}^m$ be defined by

$$Tx = (\beta_1, \dots, \beta_m), \quad x := (\alpha_1, \dots, \alpha_n) \in \mathbb{K}^n,$$

where $\beta_i := \sum_{j=1}^n a_{ij}\alpha_j$ for $i = 1, \dots, m$. If

$$a := \max_j \sum_{i=1}^m |a_{ij}|, \quad b := \max_i \sum_{j=1}^n |a_{ij}|,$$

then we can see that

$$\|Tx\|_1 \le a \|x\|_1, \quad \|Tx\|_\infty \le b \|x\|_\infty.$$

In fact, we can find unit vectors u_0, v_0 in $\mathbb{K}^n$ with $\|u_0\|_1 = 1 = \|v_0\|_\infty$ such that $\|Tu_0\|_1 = a \|u_0\|_1$ and $\|Tv_0\|_\infty = b \|v_0\|_\infty$ (see Section 3.2 in [51]). $\Diamond$

Example 2.7. Let (a_{ij}) be an infinite matrix of scalars, that is, $a_{ij} \in \mathbb{K}$ for $(i,j) \in \mathbb{N} \times \mathbb{N}$. Then, the following can be verified easily:

(i) If $a := \sup\limits_{j \in \mathbb{N}} \sum\limits_{i=1}^\infty |a_{ij}| < \infty$, then

$$\sum_{i=1}^\infty \sum_{j=1}^\infty |a_{ij}| \, |x(j)| \le a\|x\|_1 \quad \forall x \in \ell^1.$$

(ii) If $b := \sup\limits_{i \in \mathbb{N}} \sum\limits_{j=1}^\infty |a_{ij}| < \infty$, then

$$\max_{i \in \mathbb{N}} \sum_{j=1}^\infty |a_{ij}| \, |x(j)| \le b\|x\|_\infty \quad \forall x \in \ell^\infty.$$

(iii) If $c := \sum\limits_{i=1}^\infty \sum\limits_{i=1}^\infty |a_{ij}|^2 < \infty$ or if assumptions in (i) and (ii) hold, then

$$\sum_{i=1}^\infty \left(\sum_{j=1}^\infty |a_{ij}| \, |x(j)| \right)^2 \le \min\{\sqrt{ab}, \sqrt{c}\}\|x\|_2^2 \quad \forall x \in \ell^2.$$

Hence, for every $x \in \ell^\infty$,

$$(Tx)(i) := \sum_{j=1}^\infty a_{ij}x(j)$$

is well-defined for every $i \in \mathbb{N}$, and we have the following inequalities:

$$\|Tx\|_1 \leq a\|x\|_1 \quad \forall x \in \ell^1,$$
$$\|Tx\|_\infty \leq b\|x\|_\infty \quad \forall x \in \ell^\infty,$$
$$\|Tx\|_2 \leq \min\{\sqrt{ab}, \sqrt{c}\}\|x\|_2 \quad \forall x \in \ell^2.$$

Clearly, T is a linear operator on ℓ^p for $p \in \{1, 2, \infty\}$. We can also identify vectors $u_0 \in \ell^1$ and $v_0 \in \ell^\infty$ with $\|u_0\|_1 = 1 = \|v_0\|_\infty$ such that $\|Tu_0\|_1 = a\|u_0\|_1$ and $\|Tv_0\|_\infty = b\|v_0\|_\infty$. (see Section 3.2 in [51].) $\Diamond$

Example 2.8. Let $X = C[a, b]$, the space of all continuous functions on $[a, b]$ with the supremum-norm, $\|\cdot\|_\infty$. Consider $T : X \to X$ and $f : X \to \mathbb{K}$ defined by

$$(Tx)(s) = \int_a^s x(t)dt \quad \text{and} \quad f(x) = \int_a^b x(t)dt \quad \forall x \in X.$$

It is easily seen that $T \in \mathcal{B}(X)$, $f \in X'$ and $\|T\| = 1 = \|f\|$. $\Diamond$

Example 2.9. Let $k \in C([a, b] \times [a, b])$. For $x \in L^2[a, b]$, let

$$(Tx)(s) = \int_a^b k(s, t)x(t)dt, \quad a \leq s \leq b.$$

It can be seen that $Tx \in C[a, b]$ for all $x \in L^2[a, b]$. Also, we see that

$$|(Tx)(s)| \leq \|x\|_\infty \sup_{a \leq s \leq b} \int_a^b |k(s, t)|dt, \quad s \in [a, b],$$

for all $x \in C[a, b]$, and by Cauchy-Schwarz inequality,

$$|(Tx)(s)| \leq \left(\int_a^b |k(s, t)|^2 dt \right)^{1/2} \|x\|_2, \quad s \in [a, b],$$

for all $x \in L^2[a, b]$. Hence, it follows that,

$$\|Tx\|_\infty \leq \|x\|_\infty \sup_{a \leq s \leq b} \int_a^b |k(s, t)|dt \quad \forall x \in C[a, b],$$

$$\|Tx\|_\infty \leq \sup_{a \leq s \leq b} \left(\int_a^b |k(s, t)|^2 dt \right)^{1/2} \|x\|_2 \quad \forall x \in L^2[a, b],$$

$$\|Tx\|_2 \leq \left(\int_a^b \int_a^b |k(s, t)|^2 dt ds \right)^{1/2} \|x\|_2 \quad \forall x \in L^2[a, b].$$

Let $X = C[a, b]$ with $\| \cdot \|_\infty$ and $Y = L^2[a, b]$. Then, from the above inequalities, together with the fact that

$$\|x\|_2 \le (b - a)^{1/2} \|x\|_\infty \quad \forall x \in C[a, b],$$

it follows that $T \in \mathcal{B}(X)$, $T \in \mathcal{B}(X, Y)$, $T \in \mathcal{B}(Y, X)$, and $T \in \mathcal{B}(Y)$ with their norms at most equal to

$$\sup_{a \le s \le b} \int_a^b |k(s, t)| dt, \quad \left((b - a) \int_a^b \int_a^b |k(s, t)|^2 dt ds \right)^{1/2},$$

$$\sup_{a \le s \le b} \left(\int_a^b |k(s, t)|^2 dt \right)^{1/2}, \quad \left(\int_a^b \int_a^b |k(s, t)|^2 dt ds \right)^{1/2},$$

respectively. In fact,

$$\|T\|_{X \to X} = \max_{a \le s \le b} \int_a^b |k(s, t)| dt.$$

See ([51], Example 3.3 (viii)) for a proof of this. ◇

Example 2.10. For $u \in L^\infty[a, b]$, let

$$(Tx)(t) = u(t)x(t), \quad x \in L^2[a, b], \ t \in [a, b].$$

Then, for $x \in L^2[a, b]$,

$$\|Tx\|_2^2 = \int_a^b |u(t)x(t)|^2 dt \le \|u\|_\infty^2 \|x\|_2^2.$$

Thus, taking $X = L^2[a, b]$, we have $T \in \mathcal{B}(L^2[a, b])$ and $\|T\| \le \|u\|_\infty$.

If $u \in C[a, b]$, then for every $x \in C[a, b]$, we have

$$\|Tx\|_\infty \le \|u\|_\infty \|x\|_\infty.$$

Thus, taking $X = C[a, b]$ with $\| \cdot \|_\infty$, we have $T \in \mathcal{B}(X)$ and $\|T\| \le \|u\|_\infty$. In this case, it can be easily seen that $\|T\| = \|u\|_\infty$. ◇

Next we give examples of unbounded linear operators, that is, linear operators which are not continuous.

Example 2.11. Let $X = C^1[0, 1]$, the linear space of all continuously differentiable functions on $[0, 1]$ and let $Y = C[0, 1]$. Let these spaces be endowed with the same norm $\| \cdot \|_\infty$. Consider $T : X \to Y$ and $f : X \to \mathbb{K}$ defined by

$$Tx = x' \quad \text{and} \quad f(x) = x'(1), \quad x \in X.$$

It is easily seen that T and f are linear. But they are not continuous. To see this, consider

$$x_n(t) = t^n, \quad n \in \mathbb{N}.$$

Then

$$\|x_n\|_\infty = 1, \quad \|Tx_n\|_\infty = n, \quad |f(x_n)| = n \quad \forall n \in \mathbb{N}.$$

Thus $E = \{x_n : n \in \mathbb{N}\}$ is bounded in X, but its images under the maps T and f are not bounded. $\diamond$

Example 2.12. Let $X = C[0,1]$ with $\|\cdot\|_\infty$ and $Y = C[0,1]$ with $\|\cdot\|_1$. We note that

$$\|x\|_1 = \int_0^1 |x(t)|\,dt \leq \|x\|_\infty \quad \forall x \in C[0,1].$$

However, there does not exist a $c > 0$ such that the relation $\|x\|_\infty \leq c\|x\|_1$ holds for all $x \in C[0,1]$. Thus, the identity operator from X to Y is continuous. However, the identity operator from Y to X is not continuous. This is seen as follows: For $n \in \mathbb{N}$, let

$$x_n(t) = \begin{cases} n\left(\dfrac{1}{n} - t\right), & 0 \leq t \leq \frac{1}{n}, \\ 0, & \frac{1}{n} \leq t \leq 1. \end{cases}$$

Then we observe that $\|x_n\|_1 = 1/2n \to 0$ as $n \to \infty$, but $\|x_n\|_\infty = 1$ for all $n \in \mathbb{N}$. $\diamond$

Remark 2.1. In Example 2.11, we note that $N(T)$ is a one-dimensional subspace of X whereas $N(f)$ is infinite-dimensional. Not only that, $N(f)$ is not even a closed subspace. This is seen by considering

$$u_n(t) = t - t^n/n \quad \text{and} \quad u(t) = t,$$

and noting that

$$f(u_n) = 0 \quad \text{and} \quad \|u_n - u\|_\infty = 1/n \to 0 \quad \text{but} \quad f(u) = 1.$$

In fact, the non-closedness of $N(f)$ is a characteristic property of discontinuous linear functionals (see [51], Theorem 3.8). $\diamond$

A linear operator $T : X \to Y$ may not be continuous with respect to certain norms on X and Y, but can be continuous with respect to some other norms. For example, let $T : X \to Y$ be a linear operator between normed linear spaces X and Y. Then taking the norm,

$$x \mapsto \|x\|_* := \|x\|_X + \|Tx\|_Y, \quad x \in X$$

on X instead of $\| \cdot \|_X$, we see that T is continuous.

Example 2.13. In Example 2.11, if we take the norm

$$\|x\|_* := \|x\|_\infty + \|x'\|_\infty, \quad x \in C^1[0,1],$$

then it is obvious that the maps $T : x \mapsto x'$ and $f : x \mapsto x'(1)$ are continuous on X_*, the space $C^1[0,1]$ with $\| \cdot \|_*$. Now, taking

$$u_n(t) = t^n/(n+1), \quad t \in [0,1], \ n \in \mathbb{N},$$

it follows that

$$\|u_n\|_* = 1, \quad \|Tu_n\|_\infty = n/(n+1), \quad n \in \mathbb{N},$$

so that

$$\frac{n}{n+1} = \|Tu_n\|_\infty \leq \|T\| \leq 1 \quad \forall n \in \mathbb{N}.$$

Thus $\|T\| = 1$. Similarly, we see that $\|f\| = 1$. $\Diamond$

In Example 2.13, we see that there does not exist an element $x_0 \in X$ such that $\|Tx_0 = \|x_0\|$. This, in particular, shows that the closed unit ball $\{x \in X : \|x\|_* \leq 1\}$ is not compact in X_*. This is a common feature of all infinite dimensional spaces as we see in the following proposition (see [51], Theorem 2.39).

Proposition 2.1. *A closed unit ball in a normed linear space is compact if and only if the space is finite dimensional.*

The main ingredient for proving Proposition 2.1 is the following lemma (see [51], Lemma 2.40).

Lemma 2.1. *(Riesz) Suppose X is a normed linear space, X_0 is a closed subspace of X and $0 < r < 1$. Then there exists $x_0 \in X$ such that*

$$\|x_0\| = 1 \quad and \quad dist(x_0, X_0) \geq r.$$

We close this subsection by giving an important example of a linear operator, namely the projection operator which is of great importance in applications.

A linear operator $P : X \to X$ on a linear space X is called a **projection operator** or simply a **projection** if

$$Px = x \quad \forall x \in R(P).$$

Thus, a linear operator $P : X \to X$ is a projection operator if and only if $P^2 = P$.

If P is a projection, then it can be seen that $I - P$ is also a projection, and

$$R(P) = N(I - P), \quad R(I - P) = N(P).$$

Hence, if X is a normed linear space and P is continuous, then $R(P)$ and $N(P)$ are closed subspaces of X. In case X is a Banach space, we have the converse as well (see [51], Corollary 7.3). Also, if $P : X \to X$ is a nonzero continuous projection on a normed linear space, then $\|P\| \geq 1$. Indeed, for every nonzero $x \in R(P)$,

$$\|x\| = \|Px\| \leq \|P\|\,\|x\|,$$

so that $\|P\| \geq 1$.

Let X be an inner product space. Then a projection $P : X \to X$ is called an **orthogonal projection** if $R(P) \perp N(P)$, that is, if

$$\langle x, y \rangle = 0, \quad \forall x \in R(P), \quad \forall y \in N(P).$$

The following is an important observation about orthogonal projections.

Proposition 2.2. *Let $P : X \to X$ be an orthogonal projection on an inner product space X. Then $P \in \mathcal{B}(X)$ and $\|P\| = 1$.*

Proof. Since $R(I - P) = N(P) \perp R(P)$, by Pythagoras theorem 2.4, we have

$$\|x\|^2 = \|Px\|^2 + \|(I - P)x\|^2 \geq \|Px\|^2 \quad \forall x \in X.$$

Thus, $\|P\| \leq 1$. We already know that, if $P \neq 0$, then $\|P\| \geq 1$. Thus, for a nonzero orthogonal projection P, we have $\|P\| = 1$. $\qquad \square$

2.1.3 *Compact operators*

An important class of operators which occur frequently in applications is the so-called *compact operators*.

Let X and Y be normed linear spaces and $T : X \to Y$ be a linear operator. Recall that T is a bounded operator if and only if for every bounded subset E of X, the set $T(E)$ is bounded in Y so that $\mathrm{cl}T(E)$, the closure of $T(E)$, is closed and bounded. If $\mathrm{cl}T(E)$ is compact for every bounded subset E of X, then the operator T is said to be a **compact operator**. We shall denote the set of all compact operators from X to Y by $\mathcal{K}(X, Y)$. In case $Y = X$, then we denote $\mathcal{K}(X, Y)$ by $\mathcal{K}(X)$, and an operator $T \in \mathcal{K}(X)$ is said to be a **compact operator on** X.

Clearly, T is a compact operator if and only if for every bounded sequence (x_n) in X, the sequence (Tx_n) in Y has a convergent subsequence, and every compact operator is a bounded operator. By Proposition 2.1 it follows that, for any normed linear space X,

- the identity operator on X is compact if and only if X is finite dimensional, and

- a finite rank operator is compact if and only if it is a bounded operator.

Exercise 2.1. Justify the last sentence.

Suppose $T : X \to Y$ is a linear operator of *finite rank*, that is, $R(T)$ is finite dimensional. Then it can be seen that there exist vectors $v_1, \ldots, v_n$ in $R(T)$ and linear functionals $f_1, \ldots, f_n$ on X such that

$$Tx = \sum_{i=1}^{n} f_i(x)v_i, \quad x \in X. \tag{2.5}$$

This T is compact if and only if T is a bounded operator.

Exercise 2.2. Suppose $T : X \to Y$ is a finite rank operator between normed linear spaces X and Y with representation (2.5). Show that $T \in \mathcal{B}(X, Y)$ if and only if $f_i \in X'$ for every $i \in \{1, \ldots, n\}$.

Hint: Observe that $\|Tx\| \le \sum_{i=1}^{n} \|v_i\| \max\{|f_j(x)| : j = 1, \ldots, n\}$, and $\|Tx\| \ge |f_j(x)| \mathrm{dist}(v_j, X_j)$, where $X_j := \mathrm{span}\{v_i : i \ne j\}$ for each $j \in \{1, \ldots, n\}$.

One of the important examples of compact operators which is very useful in view of its applications is the *Fredholm integral operator* K defined by

$$(Kx)(s) = \int_{a}^{b} k(s,t)x(t)dt, \quad a \le s \le b, \tag{2.6}$$

where $k(\cdot, \cdot)$ is a certain admissible function on $[a, b] \times [a, b]$, and the domain and codomain of K are certain function spaces. We shall come back to this example a little later.

The following theorem lists some important properties of compact operators (see [51], Chapter 9).

Theorem 2.5. *Let X, Y and Z be normed linear spaces. Then the following hold.*

(i) *The set $\mathcal{K}(X,Y)$ is a subspace of $\mathcal{B}(X,Y)$.*

(ii) *If Y is a Banach space and if (T_n) is a sequence in $\mathcal{K}(X,Y)$ such that $\|T_n - T\| \to 0$ as $n \to \infty$ for some $T \in \mathcal{B}(X,Y)$, then $T \in \mathcal{K}(X,Y)$.*

(iii) *If $A \in \mathcal{B}(X,Y)$ and $B \in \mathcal{B}(Y,Z)$, and one of them is a compact operator, then BA is a compact operator.*

(iv) *If $T \in \mathcal{K}(X,Y)$, then $T' \in \mathcal{K}(X,Y)$.*

Corollary 2.1. *Suppose $T \in \mathcal{K}(X,Y)$. If T is bijective and $T^{-1} \in \mathcal{B}(Y,X)$, then X is finite dimensional.*

Proof. By Theorem 2.5 (iii), the operator $T^{-1}T : X \to X$ is a compact operator. But, $T^{-1}T$ is the identity operator on X which can be compact if and only if X is finite dimensional (cf. Proposition 2.1). $\qquad\square$

Example 2.14. (Diagonal operator) Suppose X and Y are Hilbert spaces, and (u_n) and (v_n) are orthonormal sequences in X and Y respectively. Suppose (λ_n) is a sequence of scalars which converges to 0. For $x \in X$, let

$$Tx := \sum_{j=1}^{\infty} \lambda_j \langle x, u_j \rangle v_j, \quad x \in X. \tag{2.7}$$

It can be seen that $T \in \mathcal{B}(X)$ and $\|T\| \leq \max_j |\lambda_j|$. Moreover, if we define

$$T_n x := \sum_{j=1}^{n} \lambda_j \langle x, u_j \rangle v_j, \quad x \in X,$$

then we see that

$$\|(T - T_n)x\|^2 := \sum_{j=n+1}^{\infty} |\lambda_j|^2 |\langle x, u_j \rangle|^2 \leq \max_{j>n} |\lambda_j|^2 \|x\|^2, \quad x \in X,$$

so that

$$\|T - T_n\| \leq \max_{j>n} |\lambda_j| \to 0 \quad \text{as} \quad n \to \infty.$$

Since each T_n is a compact operator, as it is a finite rank bounded operator, by Theorem 2.5(ii), T is also a compact operator. $\qquad\Diamond$

Remark 2.2. We shall see in Section 2.3.6 that every compact operator $T : X \to Y$ between infinite dimensional Hilbert spaces X and Y can be represented in the form (2.7); in fact, with $\lambda_j \geq 0$ for every $j \in \mathbb{N}$. $\qquad\Diamond$

Next we show that the operator K defined by (2.6) is a compact operator on certain spaces for suitable kernel function $k(\cdot,\cdot)$. For this and for many results in the sequel, we shall make use of the following result from analysis (see [51], Theorem 6.7).

Theorem 2.6. (Arzela–Ascoli) *Let Ω be a compact metric space and S be a subset of $C(\Omega)$. Then S is totally bounded with respect to the norm $\|\cdot\|_\infty$ if and only if it is pointwise bounded and equicontinuous.*

In the statement of the above theorem, we used the concepts of *pointwise boundedness, total boundedness* and *equicontinuity*. Let us recall their definitions:

Let Ω be a metric space with metric d.

(i) A subset S of Ω is said to be **totally bounded** in Ω if for every $\varepsilon > 0$, there exists a finite number of points $t_1,\ldots,t_k$ in Ω such that $S \subseteq \cup_{i=1}^k \{t \in \Omega : d(t,t_i) < \varepsilon\}$.

It is known that $S \subseteq \Omega$ is compact if and only if it is complete and totally bounded. In particular, if Ω is a complete metric space, then S is totally bounded if and only if its closure, $\mathrm{cl}(S)$, is compact.

(ii) A set S of functions from Ω to $\mathbb{K}$ is said to be **pointwise bounded** if for each $t \in S$, there exists $M_t > 0$ such that $|x(t)| \le M_t$ for all $x \in S$.

(iii) A set S of functions from Ω to $\mathbb{K}$ is said to be **equicontinuous** if for every $\varepsilon > 0$, there exists $\delta > 0$ such that
$$s,t \in \Omega, \quad d(s,t) < \delta \implies |x(t) - x(s)| < \varepsilon \quad \forall x \in S.$$

Thus, every function in an equicontinuous family is uniformly continuous.

Let us give a simple example of an equicontinuous family.

Example 2.15. For $\rho > 0$, let
$$S_\rho := \{x \in C^1[a,b] : \|x'\|_\infty \le \rho\}.$$
Then S_ρ is equicontinuous in $C[a,b]$. This follows by observing that for every $x \in S_\rho$ and $s,t \in [a,b]$,
$$x(t) - x(s) = \int_s^t x'(\tau)\, d\tau,$$
so that
$$|x(t) - x(s)| \le \rho|s - t|.$$

Note that the above set is not pointwise bounded. To see this, consider the sequence (x_n) in $C[a,b]$ defined by $x_n(t) = n$ for all $n \in \mathbb{N}$. Clearly, $x_n \in S_\rho$ for all $n \in \mathbb{N}$. However, the set

$$\widetilde{S}_\rho := \{x \in C^1[a,b] : x(a) = 0, \|x'\|_\infty \leq \rho\}$$

is both equicontinuous and pointwise bounded. Indeed, if $x \in \widetilde{S}_\rho$, then

$$x(t) = \int_a^t x'(\tau)d\tau,$$

so that $|x(t)| \leq \rho(b-a)$ for all $x \in \widetilde{S}_\rho$. Thus, in fact, $\widetilde{S}_\rho$ is uniformly bounded.

It can be easily seen that the set $\widetilde{S}_\rho$ is the image of the closed ball $\{x \in C[a,b] : \|x\|_\infty \leq \rho\}$ under the *Volterra integral operator* V defined by

$$(Vx)(s) = \int_a^s x(t)\,dt, \quad x \in C[a,b], \ s \in [a,b].$$

Thus,

$$\widetilde{S}_\rho = \{Vx : x \in C[a,b], \|x\|_\infty \leq \rho\}$$

is pointwise bounded and equicontinuous. Hence, by Arzela-Ascoli theorem (Theorem 2.6), V is a compact operator on $C[a,b]$ with respect to the norm $\|\cdot\|_\infty$. $\Diamond$

Now, we discuss the example of the operator K defined in (2.6).

Example 2.16. For $k \in C([a,b] \times [a,b])$ consider the operator K in (2.6) for $x \in C[a,b]$. We have already observed in Example 2.9 that

$$Kx \in C[a,b] \quad \forall x \in C[a,b],$$

K is a bounded operator on $C[a,b]$ with respect to the norm $\|\cdot\|_\infty$ and $\|K\| = \sup_{a \leq s \leq b} \int_a^b |k(s,t)|d$. Now, we show that K is, in fact, a compact operator. For this, first we note that, for $s, \tau \in [a,b]$ and $x \in C[a,b]$,

$$|(Kx)(s) - (Kx)(\tau)| \leq \int_a^b |k(s,t) - k(\tau,t)|\,|x(t)|\,dt$$

$$\leq \max_{a \leq t \leq b} |k(s,t) - k(\tau,t)|(b-a)\|x\|_\infty.$$

Hence, by uniform continuity of $k(\cdot,\cdot)$,

$$E := \{Kx : x \in C[a,b], \|x\|_\infty \leq 1\}$$

is a uniformly bounded and equicontinuous subset of $C[a,b]$. Therefore, by Arzela-Ascoli theorem (Theorem 2.6), the closure of E is compact in $C[a,b]$ with respect to the norm $\|\cdot\|_\infty$, and hence, K is a compact operator on $C[a,b]$ with respect to the norm $\|\cdot\|_\infty$. $\Diamond$

Remark 2.3. Using the arguments used in the above example, it can be shown that if Ω is a compact *Jordan measurable* subset of $\mathbb{R}^k$ and $k(\cdot, \cdot) \in C(\Omega \times \Omega)$, then the operator K defined by

$$(Kx)(s) = \int_\Omega k(s,t)x(t)dt, \quad s \in \Omega,$$

is a compact operator from $C(\Omega)$ to itself. Here, by Jordan measurability of Ω, we mean that the characteristic function of Ω is Riemann integrable. In fact, in this case, it is also known that

$$\|K\| = \max_{s \in \Omega} \int_\Omega |k(s,t)| \, dt.$$

For details of the above statements, one may see Kress [35]. $\Diamond$

Example 2.17. For $k \in C([a,b] \times [a,b])$ consider the integral operator K defined in (2.6) for $x \in L^2[a,b]$ with respect to the Lebesgue measure. Recall from Example 2.9 that

$$Kx \in C[a,b] \quad \forall x \in L^2[a,b].$$

For $s, \tau \in [a,b]$ and $x \in L^2[a,b]$, using Schwarz inequality, we also have

$$|(Kx)(s)| \le \int_a^b |k(s,t)x(t)|dt \le \max_{s,t \in [a,b]} |k(s,t)|(b-a)^{1/2}\|x\|_2,$$

and

$$|(Kx)(s) - (Kx)(\tau)| \le \int_a^b |k(s,t) - k(\tau,t)| \, |x(t)| \, dt$$
$$\le \max_{a \le t \le b} |k(s,t) - k(\tau,t)|(b-a)^{1/2}\|x\|_2.$$

The above two inequalities show that the set

$$E := \{Kx : x \in L^2[a,b], \|x\|_2 \le 1\}$$

is a uniformly bounded and equicontinuous subset of $C[a,b]$. Therefore, by Arzela-Ascoli theorem (Theorem 2.6), the closure of E is compact in $C[a,b]$ with respect to the norm $\|\cdot\|_\infty$. Now, to show that K is a compact operator on $L^2[a,b]$, let (x_n) be a sequence in $L^2[a,b]$ with $\|x_n\|_2 \le 1$ for all $n \in \mathbb{N}$. Then by the above observation, (Kx_n) has a convergent subsequence with respect to $\|\cdot\|_\infty$, and hence, with respect to $\|\cdot\|_2$. Thus, K is a compact operator on $L^2[a,b]$. $\Diamond$

In fact, it has been shown in [51] that (2.6) defines a compact operator from $L^p[a,b]$ to $L^r[a,b]$ for any p, r satisfying $1 \leq p \leq \infty$, $1 \leq r \leq \infty$.

Example 2.18. Let $k(\cdot, \cdot) \in L^2([a,b] \times [a,b])$ and (k_n) be a sequence in $C([a,b] \times [a,b])$ such that

$$\|k - k_n\|_2 := \int_a^b \int_a^b |k(s,t) - k_n(s,t)|^2 ds dt \to 0.$$

Let K be as in (2.6) for $x \in L^2[a,b]$ and for $n \in \mathbb{K}$, let K_n be the operator defined by

$$(K_n x)(s) = \int_a^b k_n(s,t)x(t)dt \quad x \in L^2[a,b], \quad a \leq s \leq b.$$

Then, by the results in Example 2.17 and an inequality in Example 2.9, $K_n : L^2[a,b] \to L^2[a,b]$ is a compact operator and

$$\|K - K_n\| \leq \|k - k_n\|_2 \to 0 \quad \text{as} \quad n \to \infty.$$

Therefore, by Theorem 2.5(ii), K is a compact operator on $L^2[a,b]$. $\Diamond$

Example 2.19. In this example, we take $k(\cdot, \cdot)$ to be a *weakly singular kernel* defined on $[a,b] \times [a,b]$, that is, $k(\cdot, \cdot)$ is continuous on $\{(s,t) : s \neq t\}$ and there exists β with $0 < \beta < 1$ and $M > 0$ such that

$$|k(s,t)| \leq \frac{M}{|s - t|^{1-\beta}}, \quad s \neq t.$$

Consider the integral operator K defined in (2.6) for x in either $C[a,b]$ or $L^2[a,b]$. In the following $C[a,b]$ is endowed with the norm $\|\cdot\|_\infty$. Following the arguments in Kress [35], we show that K is a compact operator on $C[a,b]$ and $L^2[a,b]$ for $\beta \in (0,1)$ and $\beta \in (1/2,1)$, respectively.

First we observe that, for any $c, d \in \mathbb{R}$ with $c < d$, the integral $\int_c^d \frac{1}{|s-t|^{1-\beta}} dt$ exists for each $s \in [c,d]$. In fact, we see that

$$\int_c^d \frac{1}{|s-t|^{1-\beta}} dt = \int_c^s \frac{1}{|s-t|^{1-\beta}} dt + \int_s^d \frac{1}{|s-t|^{1-\beta}} dt$$

$$= \frac{1}{\beta}[(s-c)^\beta + (d-s)^\beta]. \tag{2.8}$$

Let $h : [0, \infty) \to \mathbb{R}$ be defined by

$$h(t) = \begin{cases} 0, & 0 \leq t \leq 1/2, \\ 2t - 1, & 1/2 \leq t \leq 1, \\ 1, & 1 \leq t < \infty, \end{cases}$$

and for $n \in \mathbb{N}$ and $s, t \in [a, b]$, let

$$k_n(s,t) = \begin{cases} h(n|s-t|)k(s,t) & s \neq t, \\ 0, & s = t. \end{cases}$$

It can be seen that $k_n(\cdot, \cdot)$ is continuous on $[a, b] \times [a, b]$ for every $n \in \mathbb{N}$. For $x \in L^2[a, b]$, let

$$(K_n x)(s) = \int_a^b k_n(s,t)x(t)\,dt, \quad s \in [a, b].$$

Then, by the considerations in Examples 2.16 and 2.17, K_n is a compact operator on $C[a, b]$ and $L^2[a, b]$.

In order to show that K is a compact operator, by Theorem 2.5, it is enough to show that $\|K_n - K\| \to 0$ as $n \to \infty$. We show this as follows.

Observe that, if $|s - t| \geq 1/n$, then $h(n|s-t|) = 1$, so that in this case, $k_n(s,t) = k(s,t)$. Hence,

$$(Kx)(s) - (K_n x)(s) = \int_{|s-t| \leq 1/n} [k(s,t) - k_n(s,t)]x(t)\,dt.$$

Since $|k_n(s,t)| \leq |k(s,t)| \leq M/|s-t|^{1-\beta}$ for $s \neq t$, by (2.8), we have for $x \in C[a, b]$ and $s \in [a, b]$,

$$|(Kx)(s) - (K_n x)(s)| \leq 2M \int_{|s-t| \leq 1/n} \frac{|x(t)|}{|s-t|^{1-\beta}}\,dt$$

$$\leq \frac{4M}{\beta}\left(\frac{1}{n}\right)^{\beta} \|x\|_{\infty}.$$

Thus,

$$\|(K - K_n)x\|_{\infty} \leq \frac{4M}{\beta}\left(\frac{1}{n}\right)^{\beta} \|x\|_{\infty} \quad \forall x \in C[a, b],$$

so that $\|K - K_n\| \to 0$. Consequently, K is a compact operator on $C[a, b]$ for all $\beta \in (0, 1)$.

Next, for $x \in L^2[a, b]$ and $s \in [a, b]$, we have

$$|(Kx)(s) - (K_n x)(s)| \leq 2 \int_{|s-t| \leq 1/n} |k(s,t)|\,|x(t)|\,dt$$

$$\leq 2 \left(\int_{|s-t| \leq 1/n} |k(s,t)|^2\,dt \right)^{1/2} \|x\|_2$$

$$\leq 2M \left(\int_{|s-t| \leq 1/n} \frac{dt}{|s-t|^{2-2\beta}} \right)^{1/2} \|x\|_2.$$

Now, if $1/2 < \beta < 1$, then $2 - 2\beta = 1 - \mu$ with $\mu := 2\beta - 1 > 0$, so that

$$\int_{|s-t|\leq 1/n} \frac{dt}{|s-t|^{2-2\beta}} \leq \frac{2}{\mu}\left(\frac{1}{n}\right)^{\mu}.$$

Thus,

$$\|(K - K_n)x\|_2^2 \leq \frac{8M^2}{\mu}\left(\frac{1}{n}\right)^{\mu}(b-a)\|x\|_2^2 \quad \forall x \in L^2[a,b],$$

so that $\|K - K_n\| \to 0$. Consequently, K is a compact operator on $L^2[a,b]$ for $\beta \in (1/2, 1)$. ◊

Remark 2.4. It has been shown in [22] that the integral operator K defined in (2.6) is a compact operator from $L^p[a,b]$ to $C[a,b]$ (with $\|\cdot\|_\infty$) if and only if

$$\sup_{a\leq s\leq b} \int_a^b |k(s,t)|^p \, dt < \infty$$

and

$$\lim_{s\to\tau} \int_a^b |k(s,t) - k(\tau,t)|^p \, dt = 0.$$

In the above, the integral is with respect to the Lebesgue measure, and it is assumed that $k(\cdot,\cdot)$ is measurable on $[a,b] \times [a,b]$. ◊

2.2 Some Important Theorems

In this section we shall discuss some of the important theorems of basic functional analysis and their immediate consequences which are of much use in the subsequent chapters.

2.2.1 *Uniform boundedness principle*

Let X and Y be normed linear spaces and (T_n) be a sequence of operators in $\mathcal{B}(X,Y)$ which **converges pointwise on** X, that is, for each $x \in X$, $(T_n x)$ converges. Then, it can be seen easily that $T : X \to Y$ defined by for each

$$Tx := \lim_{n\to\infty} T_n x, \quad x \in X,$$

is a linear operator. However, T need not belong to $\mathcal{B}(X,Y)$ as the following example shows.

Example 2.20. Consider the vector space

$$c_{00} := \bigcup_{n=1}^{\infty} \{(\alpha_1, \ldots, \alpha_n, 0, 0, \ldots,) : \alpha_j \in \mathbb{K}, \ j = 1, 2, \ldots, n\}.$$

Let $X = c_{00}$ with the norm $\|x\|_\infty := \sup_{n \in \mathbb{N}} |x(n)|$, $x \in c_{00}$. For $n \in \mathbb{N}$, let

$$f_n(x) = \sum_{j=1}^{n} x(j), \quad x \in X.$$

Then it is easily seen that each f_n belongs to $X' = \mathcal{B}(X, \mathbb{K})$ and $\|f_n\| = n$ for all $n \in \mathbb{N}$. But the limiting operator f defined by

$$f(x) = \sum_{n=1}^{\infty} x(n), \quad x \in X,$$

does not belong to X'. This is seen by considering a sequence (x_n) defined by

$$x_n(j) = \begin{cases} 1, & j \in \{1, \ldots, n\} \\ 0, & j \notin \{1, \ldots, n\}, \end{cases}$$

and observing that $\|x_n\|_\infty = 1$ and $f(x_n) = n$ for all $n \in \mathbb{N}$. ◊

Suppose (T_n) in $\mathcal{B}(X, Y)$ converges pointwise on X. A sufficient condition to ensure the continuity of the *limiting operator* T defined by $Tx := \lim_{n \to \infty} T_n x$, $x \in X$, is the **uniform boundedness** of (T_n), that is, to have the sequence $(\|T_n\|)$ to be bounded. Indeed, if (T_n) is uniformly bounded, say $\|T_n\| \le M$ for all $n \in \mathbb{N}$, then

$$\|Tx\| = \lim_{n \to \infty} \|T_n x\| \le M \|x\| \quad \forall x \in X,$$

so that T is continuous and $\|T\| \le M$.

The above observation can be made in a more general context involving a family $\{T_\alpha\}_{\alpha \in \Lambda}$ of operators, where Λ is a subset of $\mathbb{R}$ having a limit point. For this purpose we introduce the following definition.

Let X be a normed linear space, $\Lambda \subseteq \mathbb{R}$, α_0 be a limit point of Λ and $x_\alpha \in X$ for every $\alpha \in \Lambda$. We say that (x_α) **converges to** $x \in X$ as $\alpha \to \alpha_0$, and write

$$x_\alpha \to x \quad \text{as} \quad \alpha \to \alpha_0 \quad \text{or} \quad \lim_{\alpha \to \alpha_0} x_\alpha = x,$$

if for every $\varepsilon > 0$ there exists $\delta > 0$ such that

$$\|x - x_\alpha\| < \varepsilon \quad \text{whenever} \quad \alpha \in \Lambda, \ |\alpha - \alpha_0| < \delta.$$

A family $\{T_\alpha\}_{\alpha \in \Lambda}$ of operators in $\mathcal{B}(X, Y)$ is said to be **uniformly bounded** if the set $\{\|T_\alpha\|\}_{\alpha \in \Lambda}$ is bounded in $\mathbb{R}$, and (T_α) is said to **converge pointwise** on X as $\alpha \to \alpha_0$ if $(T_\alpha x)$ converges for every $x \in X$ as $\alpha \to \alpha_0$.

The proof of the following proposition is analogous to the case of sequence of operators considered above.

Proposition 2.3. *Let X and Y be normed linear spaces, $\{T_\alpha\}_{\alpha \in \Lambda}$ be a uniformly bounded family of operators in $\mathcal{B}(X, Y)$, where Λ is a subset of $\mathbb{R}$ having a limit point α_0. If (T_α) converges pointwise on X as $\alpha \to \alpha_0$, then the operator $T : X \to Y$ defined by $Tx := \lim_{\alpha \to \alpha_0} T_\alpha x$, $x \in X$, belongs to $\mathcal{B}(X, Y)$.*

Exercise 2.3. Write the proof of the above proposition.

Next question one would like to ask is whether the convergence on the whole space X can be replaced by convergence on a dense subspace, or more generally, on a dense subset. The answer is affirmative if Y is a Banach space as the following theorem shows (see [51], Theorem 3.11).

Theorem 2.7. *Let X be a normed linear space, Y be a Banach space and (T_n) be a uniformly bounded sequence in $\mathcal{B}(X, Y)$. Suppose D is a dense subset of X such that $(T_n x)$ converges for every $x \in D$. Then $(T_n x)$ converges for every $x \in X$, and the operator $T : X \to Y$ defined by*

$$Tx = \lim_{n \to \infty} T_n x, \quad x \in X,$$

belongs to $\mathcal{B}(X, Y)$.

More generally, we have the following.

Theorem 2.8. *Let X be a normed linear space, Y be a Banach space, Λ be a subset of $\mathbb{R}$ having a limit point α_0 and $\{T_\alpha\}_{\alpha \in \Lambda}$ be a uniformly bounded family of operators in $\mathcal{B}(X, Y)$. Suppose D is a dense subset of X such that $(T_\alpha x)$ converges as $\alpha \to \alpha_0$ for every $x \in D$. Then $(T_\alpha x)$ converges as $\alpha \to \alpha_0$ for every $x \in X$ and the operator $T : X \to Y$ defined by*

$$Tx = \lim_{\alpha \to \alpha_0} T_\alpha x, \quad x \in X,$$

belongs to $\mathcal{B}(X, Y)$.

Exercise 2.4. Write the proof of the above theorem.

As we have already pointed out, pointwise convergence of (T_n) may not ensure the boundedness of $(\|T_n\|)$ (see Example 2.20). If the space X is complete, then pointwise convergence of (T_n) does imply the boundedness of $(\|T_n\|)$. This is the essence of the *Uniform Boundedness Principle* (see [51], Chapter 6).

Theorem 2.9. (Uniform Boundedness Principle) *Let X be a Banach space, Y be a normed linear space and $\{T_\alpha\}_{\alpha\in\Lambda}$ be a subset of $\mathcal{B}(X,Y)$. If $\{\|T_\alpha x\|\}_{\alpha\in\Lambda}$ is bounded for every $x \in X$, then $\{T_\alpha\}_{\alpha\in\Lambda}$ is uniformly bounded.*

Combining Theorem 2.9 and Proposition 2.3, we have the following result.

Theorem 2.10. (Banach Steinhaus Theorem) *Let X and Y be normed linear spaces and $\{T_\alpha\}_{\alpha\in\Lambda}$ be a subset of $\mathcal{B}(X,Y)$, where Λ be a subset of $\mathbb{R}$ having a limit point α_0. If X is a Banach space and $(T_\alpha x)$ converges as $\alpha \to \alpha_0$ for every $x \in X$, then $\{T_\alpha\}_{\alpha\in\Lambda}$ is uniformly bounded and the operator $T : X \to Y$ defined by*

$$Tx = \lim_{\alpha \to a_0} T_\alpha x, \quad x \in X$$

belongs to $\mathcal{B}(X,Y)$.

The following theorem, which is also known as Banach Steinhaus theorem, is a special case of Theorem 2.10.

Theorem 2.11. (Banach Steinhaus Theorem) *Let X and Y be normed linear spaces, and (T_n) be a sequence of operators in $\mathcal{B}(X,Y)$. If X is a Banach space and (T_n) converges pointwise on X, then (T_n) is uniformly bounded and the operator $T : X \to Y$ defined by*

$$Tx = \lim_{n \to \infty} T_n x, \quad x \in X,$$

belongs to $\mathcal{B}(X,Y)$.

Theorems 2.11 and 2.7 lead to the following.

Theorem 2.12. *Let X and Y be Banach spaces and (T_n) be a sequence in $\mathcal{B}(X,Y)$. Then (T_n) converges pointwise on X if and only if (T_n) is uniformly bounded and there exists a dense set $D \subseteq X$ such that $(T_n x)$ converges for each $x \in D$.*

The main nontrivial ingredient in proving the following theorem is Theorem 2.11.

Theorem 2.13. *Let X be a Banach space, Y be a normed linear spaces and (T_n) is a sequence of operators in $\mathcal{B}(X,Y)$ which converges pointwise on X. Let $T : X \to Y$ be defined by $Tx = \lim_{n\to\infty} T_n x$, $x \in X$, and $S \subseteq X$ is such that $\mathrm{cl}(S)$ is compact. Then*

$$\sup_{x\in S} \|T_n x - Tx\| \to 0 \quad \text{as} \quad n \to \infty.$$

In particular, if Z is a normed linear space and $K : Z \to X$ is a compact operator, then

$$\|(T_n - T)K\| \to 0 \quad \text{as} \quad n \to \infty.$$

Proof. By Theorem 2.11, there exists $M > 0$ such that $\|T_n\| \le M$ for every n. Let $\epsilon > 0$ be given. By the compactness of $\mathrm{cl}(S)$, there exist $x_1, \ldots, x_k$ in S such that

$$S \subseteq \bigcup_{j=1}^{k} \{x \in X : \|x - x_j\| < \epsilon\}.$$

For $i \in \{1, \ldots, k\}$, let $N_i \in \mathbb{N}$ be such that $\|T_n x_i - T x_i\| < \varepsilon$ for all $n \ge N_i$. Now, let $x \in S$, and let $j \in \{1, \ldots, k\}$ be such that $\|x - x_j\| < \varepsilon$. Then for all $n \ge N := \max\{N_i : i = 1, \ldots, k\}$, we have

$$\begin{aligned}
\|T_n x - Tx\| &\le \|T_n x - T_n x_j\| + \|T_n x_j - T x_j\| + \|T x_j - Tx\| \\
&\le \|T_n\|\|x - x_j\| + \|T_n x_j - T x_j\| + \|T\|\|x - x_j\| \\
&< (M + 1 + \|T\|)\varepsilon.
\end{aligned}$$

Since N is independent of x and ε is arbitrary, it follows that

$$\sup_{x\in S} \|T_n x - Ax\| \to 0 \quad \text{as} \quad n \to \infty.$$

The particular case follows by taking $S = \{Ku : \|u\| \le 1\}$. $\square$

Remark 2.5. We observe that, in proving Theorem 2.13, the assumption that X is a Banach space is used only to assert the boundedness of $(\|T_n\|)$, which is a consequence of Theorem 2.11. Thus, Theorem 2.13 will still hold if we replace the completeness assumption of X by the boundedness of $(\|T_n\|)$. $\Diamond$

2.2.2   *Closed graph theorem*

Among the unbounded linear operators, the so called closed linear operators are of special importance. Recall that if $T : X \to Y$ is a continuous linear operator between normed linear spaces X and Y, then for every sequence (x_n) in X with $x_n \to x$ in X, we have $Tx_n \to Tx$ in Y. In particular, the graph of the operator T, namely,

$$G(T) := \{(x, Tx) : x \in X\}$$

is a closed subspace of the product space $X \times Y$ with respect to the product norm defined by

$$\|(x, y)\| := \|x\| + \|y\|, \quad (x, y) \in X \times Y.$$

Let X and Y be normed linear spaces, X_0 be a subspace of X and let $T : X_0 \to Y$ be a linear operator. Then T is called a **closed linear operator** or a **closed operator** if the graph of T, namely,

$$G(T) := \{(x, Tx) : x \in X_0\},$$

is a closed subspace of the product space $X \times Y$.

Thus, a linear operator $T : X_0 \to Y$ is a closed linear operator if and only if for every sequence (x_n) in X_0,

$$x_n \to x \text{ in } X, \quad Tx_n \to y \text{ in } Y \implies x \in X_0, \quad Tx = y.$$

Example 2.21. Let $X = Y = C[0,1]$, $X_0 = C^1[0,1]$ and T and f be defined as in Example 2.11. It is seen that T is a closed linear operator whereas f is neither closed nor continuous.   ◇

Proposition 2.4. *Let X and Y be normed linear spaces, X_0 be a subspace of X, and $T : X_0 \to Y$ be a closed operator. Then we have the following.*

(i)   *$N(T)$ is a closed subspace of X.*
(ii) *If T is injective, then $T^{-1} : R(T) \to X$ is a closed operator.*
(iii) *If Y is a Banach space and T is a bounded operator, then X_0 is closed in X.*

Proof.   (i) Let (x_n) in $N(T)$ be such that $x_n \to x$. Since $Tx_n = 0$, by the closedness of T, we have $x \in X_0$ and $x \in N(T)$. Thus, $N(T)$ is a closed subspace.

(ii) Suppose T is injective. Let (y_n) in $R(T)$ be such that

$$y_n \to y \quad \text{and} \quad T^{-1}y_n \to x.$$

Taking $x_n = T^{-1}y_n$, it follows that $x_n \to x$ and $Tx_n \to y$ as $n \to \infty$. Hence, by the closedness of T, we have $x \in X_0$ and $Tx = y$. Thus, $y \in R(T)$ and $T^{-1}y = x$ showing that $T^{-1} : R(T) \to X$ is a closed operator.

(iii) Suppose Y is a Banach space and $T : X_0 \to Y$ is a closed operator which is also a bounded operator. Let (x_n) in X_0 be such that $x_n \to x$ in X. Since T is continuous, it follows that (Tx_n) is a Cauchy sequence in Y. Let $y = \lim_{n\to\infty} Tx_n$. Thus, we have $x_n \to x$ and $Tx_n \to y$ as $n \to \infty$, so that by the closedness of T, we have $x \in X_0$ and $Tx = y$. This shows that X_0 is closed in X. $\qquad\qquad\qquad\qquad\qquad\qquad\qquad\qquad\qquad\qquad\qquad\qquad$ $\square$

The following corollary is immediate from Theorem 2.4.

Corollary 2.2. *Let X and Y be normed linear spaces, X_0 be a subspace of X, and $T : X_0 \to Y$ be a closed operator. If X is a Banach space, T is injective and $R(T)$ is not closed in Y, then $T^{-1} : R(T) \to X$ is not continuous.*

The closed graph theorem furnishes a criterion for a closed linear operator to be continuous (cf. [51], Theorems 7.1 and 7.2).

Theorem 2.14. (Closed Graph Theorem) *Let X and Y be Banach spaces and X_0 be a subspace of X. Then a closed linear operator $T : X_0 \to Y$ is continuous if and only if X_0 is closed in X.*

The following is an immediate consequence of Theorem 2.14 and Proposition 2.4.

Theorem 2.15. (Bounded Inverse Theorem) *Let X and Y be Banach spaces, X_0 be a subspace of X and $T : X_0 \to Y$ be a closed operator. Suppose T is injective. Then $T^{-1} : R(T) \to X$ is bounded if and only if $R(T)$ is closed.*

Usually the following corollary of the above theorem is known as Bounded Inverse Theorem.

Corollary 2.3. (Bounded Inverse Theorem) *Let X and Y be Banach spaces and $T \in \mathcal{B}(X, Y)$. If T is bijective, then $T^{-1} \in \mathcal{B}(Y, X)$.*

Here is another important consequence of the closed graph theorem (see [51], Theorem 7.9).

Theorem 2.16. (Open Mapping Theorem) *Let X and Y be Banach spaces and $T \in \mathcal{B}(X, Y)$. If T is onto, then T is an open map, that is, image of every open set under T is an open set.*

Closed graph theorem also gives a criterion for continuity of projection operators (see [51], Corollary 7.3).

Corollary 2.4. *Suppose X is a Banach space and $P : X \to X$ is a projection operator. If $R(P)$ and $N(P)$ are closed subspaces of X, then P is continuous.*

The following is an important consequence of open mapping theorem (Theorem 2.16).

Theorem 2.17. *Let X and Y be Banach spaces and $T : X \to Y$ be a compact operator. If T is of infinite rank, then $R(T)$ is not closed in Y.*

Proof.   Suppose $R(T)$ is closed in Y. Then $T : X \to R(T)$ is a compact operator from the Banach space X onto the Banach space $R(T)$. Hence, by Open Mapping Theorem 2.16, there exists $\delta > 0$ such that

$$\{y \in R(T) : \|y\| \le \delta\} \subset \{Tx : \|x\| < 1\}.$$

Since the closure of the set $\{Tx : \|x\| < 1\}$ is compact, we see that the closed ball of radius δ, and thereby the closed unit ball in $R(T)$, is compact. Therefore by Proposition 2.1, $R(T)$ is finite dimensional.   □

2.2.3   Hahn-Banach theorem

Now we state one of the important results of functional analysis concerning bounded linear functionals, and consider some of its consequences (cf. [51], Chapter 5).

Recall that for a normed linear space X, X' is the dual space $\mathcal{B}(X, \mathbb{K})$.

Theorem 2.18. (Hahn-Banach Extension Theorem) *Let X_0 be a subspace of a normed linear space X and $g \in X_0'$. Then there exists $f \in X'$ such that $\|f\| = \|g\|$ and $f(x) = g(x)$ for every $x \in X_0$.*

Corollary 2.5. *If X_0 is a closed subspace of a normed linear space X and $x_0 \notin X_0$, then there exists $f \in X'$ such that $\|f\| = 1$,*

$$f(x_0) = dist(x_0, X_0) \quad and \quad f(x) = 0 \ \forall x \in X_0.$$

Taking $X_0 = \{0\}$ in the above corollary, we get

Corollary 2.6. *If x_0 is a nonzero element in a normed linear space X, then there exists $f \in X'$ such that $\|f\| = 1$ and $f(x_0) = \|x_0\|$.*

The following two corollaries are immediate consequences of the above corollary.

Corollary 2.7. *For every x in a normed linear space X,*

$$\|x\| = sup\{|f(x)| : f \in X', \|f\| \leq 1\}.$$

Corollary 2.8. *If $u_1, \ldots, u_n$ are linearly independent in a normed linear space X, then there exist linearly independent elements $f_1, \ldots, f_n$ in X' such that*

$$f_i(u_j) = \begin{cases} 1 \ if \ i = j \\ 0 \ if \ i \neq j. \end{cases}$$

As a consequence of Corollary 2.7, we have

$$\|T'\| = \|T\| \quad \forall T \in \mathcal{B}(X, Y).$$

Indeed, for $T \in \mathcal{B}(X, Y)$, we know that $\|T'\| \leq \|T\|$; and by Corollary 2.7, for every $x \in X$,

$$\begin{aligned} \|Tx\| &= sup\{|f(Tx)| : f \in X', \|f\| \leq 1\} \\ &= sup\{|(T'f)(x)| : f \in X', \|f\| \leq 1\} \\ &\leq \|T'\|\|x\| \end{aligned}$$

so that we also have $\|T'\| \leq \|T\|$.

The following result can be easily verified.

Corollary 2.9. *Let X_0 be a finite dimensional subspace of a normed linear space X, $\{u_1, \ldots, u_n\}$ be a basis of X_0 and $f_1, \ldots, f_n$ in X' be as in Corollary 2.8. Then $P : X \to X$ defined by*

$$Px = \sum_{j=1}^{n} f_j(x)x_j, \quad x \in X,$$

is a continuous projection operator with $R(P) = X_0$. In particular,

$$X = X_0 + X_1, \quad X_0 \cap X_1 = \{0\},$$

with $X_1 = N(P)$.

If the space X is an inner product space and $\{u_1, \ldots, u_n\}$ is an orthonormal basis of X_0, then Corollary 2.8 will follow without using the Hahn Banach Theorem. For, in this case, one may define $f_j : X \to X$ as

$$f_j(x) = \langle x, u_j \rangle, \quad x \in X, \quad j = 1, \ldots, n.$$

In this case, it can be seen that P is an orthogonal projection.

2.2.4   *Projection and Riesz representation theorems*

One may ask whether the conclusion of Corollary 2.9 is true if X_0 is any general closed subspace of X. The answer is, in fact, negative. For example, it is known that if $X = \ell^\infty$ and $X_0 = c_0$, the space of all sequences which converges to zero, then X_0 is a closed subspace of X but there is no closed subspace X_1 such that $X = X_0 + X_1$, $X_0 \cap X_1 = \{0\}$. In case X is a Hilbert space, then we do have a positive answer. More generally, we have the following.

Theorem 2.19. (Projection Theorem) *If X_0 is a complete subspace of an inner product space, then*

$$X = X_0 + X_0^\perp \quad and \quad (X_0^\perp)^\perp = X_0.$$

In particular, the above conclusion holds if X_0 is a closed subspace of a Hilbert space X.

Let X_0 and X be as in Theorem 2.19. Then, every $x \in X$ can be uniquely written as

$$x = u + v \quad with \quad u \in X_0, \ \ v \in X_0^\perp,$$

and in that case, $P : X \to X$ defined by $P(x) = u$, $x \in X$, is seen to be an orthogonal projection onto X_0, i.e.,

$$R(P) = X_0, \quad N(P) = X_0^\perp.$$

Corollary 2.10. *Suppose X is an inner product space and X_0 is a complete subspace of X. Then for every $x \in X$, there exists a unique element $x_0 \in X_0$ such that*

$$\|x - x_0\| = \inf\{\|x - u\| : u \in X_0\}.$$

In fact, the element x_0 is given by $x_0 = Px$, where P is the orthogonal projection of X onto X_0.

Proof. By Theorem 2.19, an orthogonal projection P onto X_0 exists. Then, for every $u \in X_0$, we have $x - u = (x - Px) + (Px - u)$, where $x - Px \in N(P)$ and $Px - u \in R(P)$. Hence, by Pythagoras theorem,

$$\|x - u\|^2 = \|x - Px\|^2 + \|Px - u\|^2$$
$$\geq \|x - Px\|^2.$$

Thus, taking $x_0 = Px$,

$$\|x - x_0\| = \inf\{\|x - u\| : u \in X_0\}.$$

If $x_1 \in X_0$ also satisfies the relation

$$\|x - x_1\| = \inf\{\|x_1 - u\| : u \in X_0\},$$

then, again by Pythagoras theorem, we have

$$\|x - x_1\|^2 = \|(x - x_0) + (x_0 - x_1)\|^2$$
$$= \|x - x_0\|^2 + \|x_0 - x_1\|^2.$$

But, $\|x - x_1\| = \|x - x_0\|$. Hence, it follows that $x_1 = x_0$, proving the uniqueness. $\qquad\square$

A subspace X_0 of a normed linear space X is said to have the **best approximation property** if for every $x \in X$, there exists $x_0 \in X_0$ such that

$$\|x - x_0\| = \inf\{\|x - u\| : u \in X_0\}.$$

If x_0 is the unique element having the above required property, then X is said to have the **unique best approximation property** with respect to X_0.

Corollary 2.10 shows that every complete subspace of an inner product space has unique best approximation property. What about for a general normed linear space?

Proposition 2.5. *Normed linear spaces have the best approximation property with respect to finite dimensional subspaces.*

Proof. Let X be a normed linear space and X_0 be a finite dimensional subspace of X. For $x \in X$, and for a fixed $v \in X_0$, consider the set

$$S = \{u \in X_0 : \|x - u\| \leq \|x - v\|\}$$

and the map

$$f : u \mapsto \|x - u\|, \quad u \in S.$$

Since the set S is closed and bounded in the finite dimensional space X_0, it is compact, and therefore, the continuous function f attains infimum at some $x_0 \in S$. Then it follows that

$$\|x - x_0\| = \inf\{\|x - u\| : u \in X_0\}.$$

This completes the proof. $\qquad\square$

It is to be remarked that the element x_0 in Proposition 2.5 need not be unique (*Exercise*).

Now we set to state another important theorem of functional analysis. To motivate the result we first recall that if X is an inner product space, then every $u \in X$ gives rise to a bounded linear functional f_u defined by $f_u(x) = \langle x, u \rangle$, $x \in X$. Clearly, $\|f_u\| = 1$. A question is whether every continuous linear functional on X is obtained in the above manner. The answer is affirmative if the space is complete.

Theorem 2.20. (Riesz Representation Theorem) *Let X be a Hilbert space. Then for every $f \in X'$, there exists a unique $u \in X$ such that*

$$f(x) = \langle x, u \rangle \quad \forall\, x \in X.$$

The above theorem need not hold in an arbitrary inner product space (cf. [51], Section 3.3).

Using Theorem 2.20, we define the **adjoint** of an operator (cf. [51], Section 3.3).

2.2.4.1 *Adjoint of an operator*

Let X and Y be inner product spaces, and let $T : X \to Y$ be a linear operator. A linear operator $T^* : Y \to X$ is called an **adjoint** of T if

$$\langle Tx, y \rangle = \langle x, T^*y \rangle, \quad \forall x \in X, y \in Y.$$

Notice that the inner products on the left-hand side and right-hand side of the above equation are that of X and Y, respectively.

If X and Y are Hilbert spaces, and $T : X_0 \to Y$ is a linear operator, where X_0 is a dense subspace of X, then the adjoint T^* of T can be defined uniquely on a dense subspace of Y as follows:

Let Y_0 be the set of all $y \in Y$ such that the linear functional $g_y : X_0 \to \mathbb{K}$ defined by

$$g_y(x) = \langle Tx, y \rangle_Y, \quad x \in X_0,$$

is continuous. Then each g_y can be extended continuously and uniquely (since X_0 is dense in X) to all of X. Therefore, by Riesz representation theorem (Theorem 2.20), there exists a unique element in X, say $T^*y \in X$, such that

$$\langle Tx, y \rangle = \langle x, T^*y \rangle \quad \forall x \in X_0.$$

The map $y \mapsto T^*y$ from Y_0 to X is the adjoint of T.

It can be verified, using the denseness of X_0, that $T^* : Y_0 \to X$ is a closed operator. Also, Y_0 is dense in Y and $T^{**} = T$, provided T is a closed operator.

It can be seen that if $T \in \mathcal{B}(X, Y)$, then $T^* \in \mathcal{B}(Y, X)$, and the following relations can be easily verified:

$$\|T^*\| = \|T\|, \quad \|T^*T\| = \|T\|^2$$

$$N(T) = R(T^*)^\perp, \quad N(T^*) = R(T)^\perp,$$

$$N(T)^\perp = \mathrm{cl}R(T^*), \quad N(T^*)^\perp = \mathrm{cl}R(T).$$

Exercise 2.5. Verify the conclusions in the above statement.

The following result will be extensively used in later chapters.

Theorem 2.21. *Let X and Y be Hilbert spaces and $T : X \to Y$ be a linear operator. Then $T \in \mathcal{K}(X, Y)$ if and only if $T^* \in \mathcal{K}(Y, X)$.*

Proof. Suppose $T \in \mathcal{K}(X, Y)$. Let (y_n) be a bounded sequence in Y. By Theorem 2.5(iii), $TT^* \in \mathcal{K}(Y)$. Hence, the sequence (TT^*y_n) has a convergent subsequence, say $(TT^*y_{n_k})$. Note that, for every $n, m \in \mathbb{N}$,

$$\|T^*y_n - T^*y_m\|^2 = \langle TT^*(y_n - y_m), (y_n - y_m) \rangle$$
$$\leq \|TT^*(y_n - y_m)\| \, \|y_n - y_m\|.$$

Since (y_n) is bounded and $(TT^*y_{n_k})$ is convergent, it follows from the above inequality that $(T^*y_{n_k})$ is a Cauchy sequence. Hence, it converges. Thus, we have proved that $T^* \in \mathcal{K}(Y, X)$. Now, the converse part is a consequence of the fact that $T = T^{**}$. □

For $T \in \mathcal{B}(X)$, where X is a Hilbert space, we define the following:

(i) T is said to be a **self adjoint operator** if $T^* = T$,
(ii) T is said to be a **normal operator** if $T^*T = TT^*$, and
(iii) T is said to be a **unitary operator** if $T^*T = I = TT^*$.

An important property of a self-adjoint operator T is that

$$\|T\| = \sup\{\langle Tx, x \rangle : \|x\| = 1\}.$$

2.3   Spectral Results for Operators

2.3.1   *Invertibility of operators*

Let X and Y be normed linear spaces and $T : X \to Y$ be a linear operator. As we have remarked in Chapter 1, it is important to know when T is bijective and its inverse T^{-1} is continuous. There are problems of practical interest in which one may know that T has continuous inverse, but one would like to know if it is the same case if T is replaced by $T + E$ for some operator $E : X \to Y$. Thus, knowing that a linear operator $T : X \to Y$ has a continuous inverse, one would like to know the conditions on $E : X \to Y$ under which $T + E$ also has continuous inverse. First let us observe the following easily verifiable result.

Proposition 2.6. *Let X and Y be normed linear spaces, X_0 be a subspace of X and $T : X_0 \to Y$ be a linear operator. Then there exists $c > 0$ such that*

$$\|Tx\| \geq c\,\|x\| \quad \forall x \in X$$

if and only if T is injective and $T^{-1} : R(T) \to Y$ is continuous, and in that case $\|T^{-1}\| \leq 1/c$.

Proof.   Left to the reader.   □

A linear operator $T : X_0 \to Y$, where X_0 is a subspace of X, is said to be **bounded below** if there exists $c > 0$ such that

$$\|Tx\| \geq c\,\|x\| \quad \forall x \in X.$$

Corollary 2.11. *Let X and Y be normed linear spaces, X_0 be a subspace of X and $T : X_0 \to Y$ be a closed linear operator. If X is a Banach space and T is bounded below, then $R(T)$ is a closed subspace of Y.*

Proof.   Suppose X is a Banach space and T is bounded below. By Proposition 2.6, $T^{-1} : R(T) \to X$ is continuous. Since $T^{-1} : R(T) \to X$ is also a closed operator, by Proposition 2.4, $R(T)$, the domain of T^{-1}, is a closed subspace of Y.   □

Corollary 2.12. *Let X be a Hilbert space and $T \in \mathcal{B}(X)$ be a self adjoint operator which is bounded below. Then T is bijective.*

Proof.   The proof follows from Corollary 2.11 by making use of the facts that $N(T)^{\perp} = \mathrm{cl}R(T^*) = \mathrm{cl}R(T)$ and $X = N(T) + N(T)^{\perp}$.   □

We shall also make use of the following two results (see [51], Theorem 10.8 and Corollary 10.9) to derive a perturbation result which will be used subsequently.

We recall from bounded inverse theorem (Corollary 2.3) that if X and Y are Banach spaces and $T \in \mathcal{B}(X, Y)$ is bijective then $T^{-1} \in \mathcal{B}(Y, X)$.

The following result is crucial for obtaining perturbation results (see [51], Section 10.2).

Theorem 2.22. *Let X be a Banach space and $A \in \mathcal{B}(X)$. If $\|A\| < 1$, then $I - A$ is bijective and*

$$\|(I - A)^{-1}\| \le \frac{1}{1 - \|A\|}.$$

As a consequence of the above result, we have the following perturbation result.

Corollary 2.13. *Let X be a normed linear space, Y be a Banach space and $T : X \to Y$ be a bijective linear operator with $T^{-1} \in \mathcal{B}(Y, X)$. If $E : X \to Y$ is a linear operator such that $ET^{-1} \in \mathcal{B}(Y)$ and $\|ET^{-1}\| < 1$, then $T + E$ is bijective, $(T + E)^{-1} \in \mathcal{B}(Y, X)$ and*

$$\|(T + E)^{-1}\| \le \frac{\|T^{-1}\|}{1 - \|ET^{-1}\|}.$$

Proof. Since $\|ET^{-1}\| < 1$, by Theorem 2.22 with $A = ET^{-1}$, the operator $I + ET^{-1}$ on the Banach space Y is bijective and

$$\|(I + ET^{-1})^{-1}\| \le \frac{1}{1 - \|ET^{-1}\|}.$$

Hence, $T + E = (I + ET^{-1})T$ is also bijective and

$$(T + E)^{-1} = T^{-1}(I + ET^{-1})^{-1}.$$

From this it follows that $(T + E)^{-1} \in \mathcal{B}(Y, X)$ so that

$$\|(T + E)^{-1}\| \le \frac{\|T^{-1}\|}{1 - \|ET^{-1}\|}.$$

This completes the proof. □

A more general result than the above is possible by making use of the following corollary.

Corollary 2.14. *Let X be a Banach space and $A \in \mathcal{B}(X)$. If $\|A^k\| < 1$ for some positive integer k, then $I - A$ is bijective and*

$$\|(I - A)^{-1}\| \le \frac{\|\sum_{i=0}^{k-1} A^i\|}{1 - \|A^k\|}.$$

Proof. Using the identity

$$I - A^k = (I - A)(I + A + \ldots + A^{k-1})$$
$$= (I + A + \ldots + A^{k-1})(I - A),$$

the proof can be deduced from Theorem 2.22. $\square$

The following perturbation result is under weaker assumption than in Corollary 2.13.

Corollary 2.15. *Let X be a normed linear space, Y be a Banach space and $T : X \to Y$ be a bijective linear operator with $T^{-1} \in \mathcal{B}(Y, X)$. If $E : X \to Y$ is a linear operator such that $ET^{-1} \in \mathcal{B}(Y)$ and $\|(ET^{-1})^2\| < 1$, then $T + E$ is bijective, $(T + E)^{-1} \in \mathcal{B}(Y, X)$ and*

$$\|(T + E)^{-1}\| \leq \frac{\|T^{-1}\|(1 + \|ET^{-1}\|)}{1 - \|(ET^{-1})^2\|}.$$

Proof. Since $\|(ET^{-1})^2\| < 1$, Corollary 2.14 with $A = ET^{-1}$ and $k = 2$ implies that the operator $I + ET^{-1}$ on the Banach space Y is bijective, and

$$\|(I + ET^{-1})^{-1}\| \leq \frac{1 + \|ET^{-1}\|}{1 - \|(ET^{-1})^2\|}.$$

Therefore, $T + E = (I + ET^{-1})T$ is bijective and

$$(T + E)^{-1} = T^{-1}(I + ET^{-1})^{-1}$$

so that $(T + E)^{-1} \in \mathcal{B}(Y, X)$ and

$$\|(T + E)^{-1}\| \leq \frac{\|T^{-1}\|(1 + \|ET^{-1}\|)}{1 - \|(ET^{-1})^2\|}.$$

This completes the proof. $\square$

2.3.2   *Spectral notions*

An important set of scalars associated with a linear operator is its *spectrum*. Let X_0 be a subspace of a normed linear space X and let $A : X_0 \to X$ be a linear operator. Then the set

$$\rho(A) := \{\lambda \in \mathbb{K} : A - \lambda I : X_0 \to X \text{ bijective and } (A - \lambda I)^{-1} \in \mathcal{B}(X)\}$$

is called the **resolvent set** of A, and its compliment in $\mathbb{K}$,

$$\sigma(A) := \{\lambda \in \mathbb{K} : \lambda \notin \rho(A)\},$$

is called the **spectrum** of A. Elements of $\sigma(A)$ are called **spectral values** of A, and the quantity

$$r_\sigma(A) := \sup\{|\lambda| : \lambda \in \sigma(A)\}$$

is called the **spectral radius** of A.

In view of the bounded inverse theorem (Theorem 2.15), if X is a Banach space and $A : X_0 \to X$ is a closed operator, then

$$\rho(A) = \{\lambda \in \mathbb{K} : A - \lambda I : X_0 \to X \text{ is bijective}\}.$$

The set of all scalars λ for which the operator $A - \lambda I$ is not injective is called the **eigen spectrum** of A, and it is denoted by $\sigma_e(A)$. Thus,

$$\sigma_e(A) \subseteq \sigma(A).$$

Elements of the eigen spectrum are called **eigenvalues**. Thus, a scalar λ is an eigenvalue of A if and only if there exists a nonzero $x \in X_0$ such that

$$Ax = \lambda x,$$

and in that case x is called an **eigen vector** of A corresponding to the eigenvalue λ. The set of all eigen vectors corresponding to an eigenvalue λ together with the zero vector, that is, the subspace $N(A - \lambda I)$, is called the **eigen space** of A corresponding to the eigenvalue λ.

In the case of a finite dimensional X, we know that every linear operator $A : X \to X$ is continuous, and A injective if and only if it is bijective, so that, in this case, we have

$$\sigma(A) = \sigma_e(A).$$

We may also observe that if $\mathbf{A}$ is an $n \times n$ matrix of scalars, then it can be considered as a linear operator $A : \mathbb{K}^n \to \mathbb{K}^n$ by defining $Ax = \mathbf{A}\mathbf{x}$ for $x \in \mathbb{K}^n$, where $\mathbf{x}$ denotes the column vector obtained by transposing $x \in \mathbb{K}^n$. Then it can be seen that

$$\lambda \in \sigma_e(A) \iff \exists \text{ nonzero } x \in \mathbb{K}^n \text{ such that } \mathbf{A}\mathbf{x} = \lambda \mathbf{x}$$
$$\iff det(\mathbf{A} - \lambda \mathbf{I}) = 0.$$

If $\lambda \in \mathbb{K}$ is such that $A - \lambda I$ is not bounded below, then we say that λ is an **approximate eigenvalue** of A, and the set of all approximate eigenvalues of A is denoted by $\sigma_a(A)$. It can be shown that $\lambda \in \sigma_a(A)$ if and only if there exists (x_n) in X_0 such that $\|x_n\| = 1$ for all $n \in \mathbb{N}$ and

$$\|Ax_n - \lambda x_n\| \to 0 \quad \text{as} \quad n \to \infty.$$

Clearly,

$$\sigma_e(A) \subseteq \sigma_a(A) \subseteq \sigma(A).$$

If X is a Banach space and A is a compact operator on X of infinite rank, then we know (cf. Theorem 2.17) that $R(A)$ is not closed in X, so that by Corollary 2.11, $0 \in \sigma_a(A)$.

The following result can be verified easily.

Theorem 2.23. *Let X and Y be Banach spaces, and $A \in \mathcal{B}(X,Y)$, $B \in \mathcal{B}(Y,X)$ and λ be a nonzero scalar. Then*

$$\lambda \in \rho(AB) \iff \lambda \in \rho(BA),$$

and in that case

$$(BA - \lambda I)^{-1} = \frac{1}{\lambda}[B(AB - \lambda I)^{-1}A - I].$$

In particular,

$$\sigma(AB) \setminus \{0\} = \sigma(BA) \setminus \{0\} \quad and \quad r_\sigma(AB) = r_\sigma(BA).$$

It can be verified that if X is a Hilbert space and $A \in \mathcal{B}(X)$, then

$$\sigma_a(A) \subseteq \mathrm{cl}\{\langle Ax, x \rangle : \|x\| = 1\},$$

and if A is a normal operator, then

$$\lambda \in \sigma_e(A) \iff \overline{\lambda} \in \sigma_e(A^*).$$

Exercise 2.6. Prove the above facts.

The set

$$w(A) := \{\langle Ax, x \rangle : x \in X, \|x\| = 1\}$$

is called the **numerical range** of A.

The following result is significant (see [51], Section 12.1).

Theorem 2.24. *Let X be a Hilbert space and $A \in \mathcal{B}(X)$. Then*

$$\sigma(A) = \sigma_a(A) \cup \{\overline{\lambda} : \lambda \in \sigma_e(A^*)\}.$$

In particular,

(i) $\sigma(A) \subseteq \mathrm{cl}\, w(A)$,
(ii) *if A is a normal operator, then $\sigma(A) = \sigma_a(A)$, and*
(iii) *if A is self adjoint, then $\sigma(A) = \sigma_a(A) \subseteq \mathbb{R}$.*

Now, let X be a Hilbert space and $A \in \mathcal{B}(X)$. From the above theorem, it is clear that

$$r_\sigma(A) \leq r_w(A) \leq \|A\|,$$

where

$$r_w(A) := \sup\{|\lambda| : \lambda \in \omega(A)\},$$

called the **numerical radius** of A. Further, if A is a self adjoint operator, then

$$\sigma(A) \subseteq [\alpha_A, \beta_A],$$

where

$$\alpha_A = \inf w(A), \quad \beta_A = \sup w(A).$$

An operator $A \in \mathcal{B}(X)$ is called a **positive operator** if

$$w(A) \subseteq [0, \infty).$$

It is also known (cf. [51], Theorem 12.8) that for a self adjoint operator $A \in \mathcal{B}(X)$,

$$r_\sigma(A) = \|A\| = \sup\{|\langle Ax, x\rangle| : x \in X, \|x\| = 1\}. \tag{2.9}$$

By this relation it follows that the spectrum of a self adjoint operator is non-empty.

We observe that if X and Y are Hilbert spaces and $T \in \mathcal{B}(X, Y)$, then the operators $T^*T \in \mathcal{B}(X)$ and $TT^* \in \mathcal{B}(Y)$ are positive as well as self adjoint.

In general, we have the following (see [51], Section 10.2).

Proposition 2.7. *Let X be a Banach space and $A \in \mathcal{B}(X)$. Then $\rho(A)$ is open, $\sigma(A)$ is compact and*

$$r_\sigma(A) \leq \inf\{\|A^k\|^{1/k} : k \in \mathbb{N}\}.$$

In case $\mathbb{K} = \mathbb{C}$, then $\sigma(A)$ is nonempty, and

$$r_\sigma(A) = \inf\{\|A^k\|^{1/k} : k \in \mathbb{N}\} = \lim_{k \to \infty} \|A^k\|^{1/k}. \tag{2.10}$$

The equality in (2.10) need not hold if $\mathbb{K} = \mathbb{R}$. For example, if we take $X = \mathbb{R}^2$ and A is the operator which maps $x := (\alpha_1, \alpha_2)$ to $Ax := (\alpha_2, -\alpha_1)$, then it is seen that $\sigma(A) = \sigma_e(A) = \varnothing$. Since

$$A^{2k} = (-I)^k, \quad A^{2k+1} = (-I)^k A \quad \forall k \in \mathbb{N},$$

for any norm on $\mathbb{R}^2$, we have

$$0 = r_\sigma(A) < \min\{1, \|A\|\} = \inf\{\|A^k\|^{1/k} : k \in \mathbb{N}\}.$$

Example 2.22. Let $X = C^1[0,1]$ and let $A : X \to X$ be defined by $Ax = x'$, the derivative of x. Then it can be easily seen that for $\lambda \in \mathbb{K}$, a function $x \in X$ satisfies the equation $Ax = \lambda x$ if and only if $x(t) = e^{\lambda t}$, $t \in [0,1]$. Thus, $\sigma_e(A) = \mathbb{K}$. If we take

$$X_0 := \{x \in C^1[0,1] : x(0) = 0\},$$

and consider A as an operator on X_0, then we have $\sigma_e(A) = \varnothing$. $\Diamond$

Example 2.23. Let $X = C[0,1]$ with $\|\cdot\|_\infty$ and let $A : X \to X$ be defined by

$$(Ax)(t) = \int_0^t x(s)ds, \quad x \in C[0,1].$$

Clearly for $x \in X$, $Ax = 0$ if and only if $x = 0$ so that $0 \notin \sigma_e(A)$. Now, let λ be a nonzero scalar and $x \in X$. Then we see that $Ax = \lambda x$ if and only if

$$x(t) = \frac{1}{\lambda} \int_0^t x(s)ds.$$

There does not exist a nonzero function x in $C[0,1]$ satisfying the above equation, as the above equation for a nonzero x implies that x is differentiable, $x(0) = 0$ and $x(t) = e^{t/\lambda}$ for all $t \in [0,1]$, which is impossible. Thus, $\sigma_e(A) = \varnothing$. One may ask whether A has a nonzero spectral value. Recall that A is a compact operator. We shall see that every nonzero spectral value of a compact operator is an eigenvalue. Hence, for the above A, $\sigma(A) \subseteq \{0\}$. Since A cannot be surjective, as every function in the range of A has to be differentiable, it follows that $0 \in \sigma(A)$. Thus, we have $\sigma(A) = \{0\}$, and 0 is not an eigenvalue. $\Diamond$

For more examples of eigenvalues, approximate eigenvalues, spectral values, and spectral properties of operators, one may refer ([51], Chapters 10, 12, 13).

2.3.3   *Spectrum of a compact operator*

We consider some important results regarding the spectral values of compact operators on a Banach space and self adjoint operators on a Hilbert space (cf. [51], Chapter 9).

Proposition 2.8. *Let X be a Banach space, $A \in \mathcal{B}(X)$ be a compact operator and $0 \neq \lambda \in \mathbb{K}$. Then*

(i) $N(A - \lambda I)$ *is finite dimensional, and*

(ii) $R(A - \lambda I)$ *is closed.*

Theorem 2.25. *Let X be a Banach space and $A \in \mathcal{B}(X)$ be a compact operator. Then we have the following.*

(i) *Every nonzero spectral value of A is an eigenvalue of A, and it is an isolated point of $\sigma(A)$.*

(ii) $\sigma(A)$ *is countable.*

Theorem 2.25 together with the relation (2.9) shows that every nonzero compact self adjoint operator on a Hilbert space has a nonzero eigenvalue.

2.3.4 *Spectral Mapping Theorem*

Let X be a Banach space and $A \in \mathcal{B}(X)$. If $p(t)$ is a polynomial, say $p(t) = a_0 + a_1 t + \ldots + a_n t^n$, then we define

$$p(A) = a_0 I + a_1 A + \ldots + a_n A^n.$$

One may enquire how the spectrum of $p(A)$ is related to the spectrum of A. In this connection we have the following theorem (see [51], Theorems 10.14 and 12.12).

Theorem 2.26. *Let X be a Banach space, $A \in \mathcal{B}(X)$ and $p(t)$ be a polynomial. Then*

$$\{p(\lambda) : \lambda \in \sigma(A)\} \subseteq \sigma(p(A)).$$

If either $\mathbb{K} = \mathbb{C}$ or if X is a Hilbert space and A is self adjoint, then

$$\{p(\lambda) : \lambda \in \sigma(A)\} = \sigma(p(A)).$$

Corollary 2.16. *Let X be a Hilbert space, $A \in \mathcal{B}(X)$ be a self adjoint operator and $p(t)$ be a polynomial with real coefficients. Then $p(A)$ is self adjoint and*

$$\|p(A)\| \leq \|p\|_\infty := \sup\{|p(\lambda)| : \lambda \in [\alpha_A, \beta_A]\},$$

where $\alpha_A := \inf w(A)$ and $\beta_A := \sup w(A)$.

Now, suppose that $A \in \mathcal{B}(X)$ is a self adjoint operator on a Hilbert space X and f is a real valued continuous function defined on $[\alpha_A, \beta_A]$. By Weierstrass approximation theorem, there exists a sequence (p_n) of polynomials with real coefficients such that

$$\|f - p_n\|_\infty := \sup_{\alpha_A \leq t \leq \beta_A} |f(t) - p_n(t)| \to 0$$

as $n, m \to \infty$. By Corollary 2.16,

$$\|p_n(A) - p_m(A)\| \le \|p_n - p_m\|_\infty.$$

Hence, by completeness of $\mathcal{B}(X)$, the sequence $(p_n(A))$ converges to an operator in $\mathcal{B}(X)$. We denote this operator by $f(A)$, that is,

$$f(A) := \lim_{n \to \infty} p_n(A). \tag{2.11}$$

It can be easily seen that $f(A)$ is also self adjoint. Moreover, by Corollary 2.16,

$$\|f(A)\| = \lim_{n \to \infty} \|p_n(A)\| \le \lim_{n \to \infty} \|p_n\|_\infty = \|f\|_\infty. \tag{2.12}$$

If A is a positive self adjoint operator, then using the definition (2.11), we can define A^ν, $\nu > 0$, by taking $f(\lambda) = \lambda^\nu$; in particular, square root of A, $A^{1/2}$, is the operator $f(A)$ with $f(\lambda) = \sqrt{\lambda}$, $\lambda > 0$.

2.3.5　*Spectral representation for compact self adjoint operators*

We know from linear algebra that if A is a self adjoint operator on a finite dimensional inner product space, then it can be represented in terms of its eigenvalues and a basis consisting of orthonormal eigen vectors. This result can be extended to any compact self adjoint operator on a Hilbert space as follows.

Theorem 2.27. *Let X be a Hilbert space and $A \in \mathcal{B}(X)$ be a compact self adjoint operator. Let $\sigma(A) = \{\lambda_j : j \in \Lambda\}$ where Λ is either $\{1, 2, \dots, n\}$ for some $n \in \mathbb{N}$ or $\Lambda = \mathbb{N}$ according as $\sigma(A)$ is finite or infinite. For each nonzero eigenvalue λ_j, let $\{u_1^{(j)}, \dots, u_{m_j}^{(j)}\}$ be an orthonormal basis of $N(A - \lambda_j I)$ and P_j be the orthogonal projection onto $N(A - \lambda_j I)$. Then*

$$Ax = \sum_{j \in \Lambda} \sum_{i=1}^{m_j} \lambda_j \langle x, u_i^{(j)} \rangle u_i^{(j)} \quad \forall x \in X,$$

and

$$A = \sum_{j \in \Lambda} \lambda_j P_j.$$

Moreover $\cup_{j \in \Lambda} \{\{u_1^{(j)}, \dots, u_{m_j}^{(j)}\}$ is an orthonormal basis of $N(A)^\perp$.

From Theorem 2.27 we see that if A is a compact self adjoint operator on a Hilbert space then there exists a sequence (μ_n) of scalars and an orthonormal set $\{u_n : n \in \mathbb{N}\}$ in X satisfying

$$Ax = \sum_{n=1}^{\infty} \mu_n \langle x, u_n \rangle u_n \quad \forall\, x \in X. \tag{2.13}$$

From the above representation of A, it follows that, for every polynomial p,

$$p(A)x = \sum_{n=1}^{\infty} p(\mu_n) \langle x, u_n \rangle u_n, \quad x \in X,$$

and hence, using the definition (2.11), it can be verified that for every continuous real valued function f on $[a, b]$,

$$f(A)x = \sum_{n=1}^{\infty} f(\mu_n) \langle x, u_n \rangle u_n, \quad x \in X. \tag{2.14}$$

Exercise 2.7. Let A be as in (2.13). Then, for every continuous real valued function f defined on $[a, b] \supseteq \mathrm{cl}\{\mu_n : n \in \mathbb{N}\}$, $f(A)$ given in (2.14) is a compact self adjoint operator. Why?

2.3.6 *Singular value representation*

One may ask whether a representation similar to (2.13) is possible if A is a compact operator which is not a self adjoint. There are compact operators having no nonzero eigenvalues. Therefore, a representation as in (2.13), in terms of eigenvalues, is not possible for a general compact operator. But we do obtain a representation in terms of *singular values* of A.

We have observed in Theorem 2.5 that if $A, B \in \mathcal{B}(X)$ and one of them is compact, then their products, AB and BA, are also compact operators. Therefore, for any compact operator T on a Hilbert space X, the operators T^*T and TT^* are also compact. Moreover they are self adjoint operators. Since T^*T is also a positive operator, there exist non-negative scalars σ_n and orthonormal basis $\{u_n : n \in \mathbb{N}\}$ for $N(T^*T)^\perp$ such that

$$T^*Tx = \sum_{n=1}^{\infty} \sigma_n^2 \langle x, u_n \rangle u_n \quad \forall\, x \in X. \tag{2.15}$$

Note that

$$T^*Tu_n = \sigma_n^2 u_n, \quad n \in \mathbb{N}.$$

The scalars σ_n, $n \in \mathbb{N}$, are called the **singular values** of the compact operator T. Let $v_n = Tu_n/\sigma_n$ for $n \in \mathbb{N}$. Then we see that

$$Tu_n = \sigma_n v_n \quad \text{and} \quad T^* v_n = \sigma_n u_n$$

for every $n = 1, 2, \ldots$. The set $\{(\sigma_n, u_n, v_n) : n \in \mathbb{N}\}$ is called a **singular system** for the compact operator T.

Theorem 2.28. *Let $T : X \to Y$ be a compact operator between Hilbert spaces X and Y and $\{(\sigma_n, u_n, v_n) : n \in \mathbb{N}\}$ be a singular system for T. Then $\{u_n : n \in \mathbb{N}\}$ and $\{v_n : n \in \mathbb{N}\}$ are orthonormal bases of $N(T)^\perp$ and $clR(T)$ respectively, and for $x \in X$ and $y \in Y$,*

$$Tx = \sum_{n=1}^{\infty} \sigma_n \langle x, u_n \rangle v_n, \quad T^* y = \sum_n \sigma_n \langle y, v_n \rangle u_n. \tag{2.16}$$

Proof. Recall that $\{u_n : n \in \mathbb{N}\}$ is an orthonormal basis of $N(T^*T)^\perp$. Also, we have $N(T^*T)^\perp = N(T)^\perp$. Thus $\{u_n : n \in \mathbb{N}\}$ is an orthonormal basis of $N(T)^\perp$.

Now, to show that $\{v_n : n \in \mathbb{N}\}$ is an orthonormal basis of $clR(T)$, it is enough (*Why ?*) to show that

$$y \in R(T), \ \langle y, v_n \rangle = 0 \ \forall n \in \mathbb{N} \implies y = 0.$$

So, let $y \in R(T)$ such that $\langle y, v_n \rangle = 0$ for all $n \in \mathbb{N}$. Let $x \in N(T)^\perp$ be such that $y = Tx$. Then, for all $n \in \mathbb{N}$, we have

$$\langle y, v_n \rangle = \langle Tx, v_n \rangle = \langle x, T^* v_n \rangle = \langle x, \sigma_n u_n \rangle = \sigma_n \langle x, u_n \rangle.$$

Thus, $\langle x, u_n \rangle = 0$ for every $n \in \mathbb{N}$. Since $\{u_n : n \in \mathbb{N}\}$ is an orthonormal basis of $N(T)^\perp$, it follows that $x = 0$ so that $y = 0$. Since $\{v_n : n \in \mathbb{N}\}$ is an orthonormal basis of $clR(T)$, and since for $x \in X$,

$$\langle Tx, v_n \rangle = \langle x, T^* v_n \rangle = \sigma_n \langle x, u_n \rangle \quad \forall n \in \mathbb{N}$$

we have, by Fourier expansion (Theorem 2.3),

$$Tx = \sum_{n=1}^{\infty} \langle Tx, v_n \rangle v_n = \sum_{n=1}^{\infty} \sigma_n \langle x, u_n \rangle v_n.$$

Also, since $\{u_n : n \in \mathbb{N}\}$ is an orthonormal basis of $N(T)^\perp$ and $R(T^*) \subseteq N(T)^\perp$, for $y \in Y$, we have

$$T^* y = \sum_{n=1}^{\infty} \langle T^* y, u_n \rangle v_n = \sum_{n=1}^{\infty} \sigma_n \langle y, v_n \rangle u_n.$$

This completes the proof. □

The representation of T in (2.16) is called the **singular value representation** of T.

If X and Y are Euclidean spaces, then the representation of T in (2.16) is nothing but the singular value decomposition of T.

2.3.7   *Spectral theorem for self adjoint operators*

Now we take up the case when $A \in \mathcal{B}(X)$ is a self adjoint operator, but not necessarily a compact operator. In this case also we have a representation theorem, but not in terms of sums but in terms of integrals (cf. [51], Section 13.3).

As earlier, for a self adjoint operator A, let us denote

$$\alpha_A = \inf w(A) \quad \text{and} \quad \beta_A = \sup w(A),$$

where $w(A)$ is the numerical range of A.

Theorem 2.29. *Let X be a Hilbert space and $A \in \mathcal{B}(X)$ be a self adjoint operator. Then there exists a family $\{E_\lambda : a \leq \lambda \leq b\}$ of orthogonal projections with $a = \alpha_A$ and $b = \beta_A$ such that $E_a = O$, $E_b = I$, $E_\lambda \leq E_\mu$ whenever $a \leq \lambda \leq \mu \leq b$, $E_{\lambda+1/n} x \to E_\lambda x$ as $n \to \infty$ for every $x \in X$ and $a < \lambda < b$, and*

$$A = \int_a^b \lambda \, dE_\lambda. \tag{2.17}$$

Moreover, for every real valued $f \in C[a,b]$,

$$f(A) := \int_a^b f(\lambda) \, dE_\lambda \tag{2.18}$$

is a self adjoint operator on X, and

$$\langle f(A)x, y \rangle := \int_a^b f(\lambda) \, d\langle E_\lambda x, y \rangle \quad \forall x, y \in X. \tag{2.19}$$

In particular,

$$\|f(A)x\|^2 = \int_a^b |f(\lambda)|^2 d\langle E_\lambda x, x \rangle \quad \forall x \in X. \tag{2.20}$$

The family $\{E_\lambda : \alpha_A \leq \lambda \leq \beta_A\}$ of projections in Theorem 2.29 is called the **normalized resolution of identity** corresponding to A, and the integral representation (2.17) of A is called the **spectral representation** of A. The integrals in (2.18) and (2.19) are in the sense of Riemann-Stieltjes. For instance, $f(A)$ in (2.18) is defined as follows: For every $\varepsilon > 0$, there exists a $\delta > 0$ such that

$$\left\| f(A) - \sum_{j=1}^n f(\tau_j^{(n)})(E_{\lambda_j^{(n)}} - E_{\lambda_{j-1}^{(n)}}) \right\| < \varepsilon$$

whenever

$$\alpha_A = \lambda_0^{(n)} \leq \lambda_1^{(n)} \leq \ldots \leq \lambda_n^{(n)} = \beta_A, \quad \tau_j^{(n)} \in [\lambda_{j-1}^{(n)}, \lambda_j^{(n)}], \quad j = 1, \ldots, n$$

and

$$\max_{1 \le j \le n} (\lambda_j^{(n)} - \lambda_{j-1}^{(n)}) < \delta.$$

It can be shown that the operator $f(A)$ defined above is the same as the one defined in (2.11), so that

$$\|f(A)\| \le \sup\{|f(\lambda)| : \alpha_A \le \lambda \le \beta_A\}. \tag{2.21}$$

We end this chapter here. Additional results from functional analysis and basic operator theory will be recalled from standard texts as and when necessary.

PROBLEMS

(1) Let X be a linear space and Y be a normed linear space. If $T : X \to Y$ is an injective linear operator, then show that $\|x\| := \|Tx\|_Y$, $x \in X$, defines a norm on X.

(2) Let $\|\cdot\|$ and $\|\cdot\|_*$ be norms on a linear space X such that there exist $c_1 > 0$, $c_2 > 0$ such that $c_1\|x\| \le \|x\|_* \le c_2\|x\|$ for all $x \in X$. Show that X is a Banach space with respect to $\|\cdot\|$ if and only if X is a Banach space with respect to $\|\cdot\|_*$ as well.

(3) Let X be an inner product space and $S \subseteq X$. Show that $S^\perp$ is a closed subspace of X and $S \cap S^\perp \subseteq \{0\}$.

(4) Show that every orthogonal set which does not contain the zero vector is linearly independent.

(5) Show that an orthonormal subset E of an inner product space is an orthonormal basis if and only if $E^\perp = \{0\}$.

(6) Suppose X and Y are normed linear spaces, $T : X \to Y$ be a linear operator and $S := \{x \in X : \|x\| = 1\}$. Show that T is continuous if and only if $F : S \to Y$ defined by $F(x) = Tx$, $x \in S$, is a bounded function.

(7) Let X and Y be normed linear spaces and $T : X \to Y$ be a linear operator. If T is an open map, then show that T is onto.

(8) Let X and Y be normed linear spaces and $T : X \to Y$ be a linear operator. If there exist $c \ge 0$ and a nonzero vector $x_0 \in X$ such that $\|Tx\| \le c\|x\|$ for all $x \in X$ and $\|Tx_0\| = c\|x_0\|$, then show that $T \in \mathcal{B}(X, Y)$ and $\|T\| = c$.

(9) Let X be an inner product space, $u \in X$ and $f : X \to \mathbb{K}$ be defined by $f(x) = \langle x, u \rangle$, $x \in X$. Show that f is a continuous linear functional and $\|f\| = \|u\|$.

(10) Let (a_{ij}) be an $m \times n$ matrix of scalars and let $T : \mathbb{K}^n \to \mathbb{K}^m$ be defined by

$$(Tx)(i) = \sum_{j=1}^{n} a_{ij}x(j), \quad x \in \mathbb{K}^n, \quad i = 1, \ldots, m.$$

Let $a := \max_j \sum_{i=1}^{m} |a_{ij}|$ and $b := \max_i \sum_{j=1}^{n} |a_{ij}|$. Find nonzero vectors u_0, v_0 in $\mathbb{K}^n$ such that $\|u_0\|_1 = 1 = \|v_0\|_\infty$ and $\|Tu_0\|_1 = a\|u_0\|_1$ and $\|Tv_0\|_\infty = b\|v_0\|_\infty$.

(11) Let T and f be as in Example 2.8. Find nonzero u and v in $C[a, b]$ such that $\|Tu\| = \|u\|_\infty$ and $|f(v)| = \|v\|_\infty$.

(12) Show that there does not exist a constant $c > 0$ such that $\|x\|_\infty \leq c\|x\|_2$ for all $x \in C[a, b]$.

(13) If the space X is an inner product space and $\{u_1, \ldots, u_n\}$ is an orthonormal subset of X, then show that $P : X \to X$ defined by $Px = \sum_{j=1}^{n}\langle x, u_j \rangle u_j$, $x \in X$, is an orthogonal projection with $R(P) = \text{span}\{u_1, \ldots, u_n\}$.

(14) Let X and Y be inner product spaces and $A : X \to Y$ and $B : Y \to X$ be linear operators such that $\langle Ax, y \rangle = \langle x, By \rangle$ for all $x \in X$, $y \in Y$. Show that $A \in \mathcal{B}(X, Y)$ if and only if $B \in \mathcal{B}(Y, X)$.

(15) Let X and Y be Hilbert spaces and $T : X \to Y$ be a linear operator such that $\langle Tx, y \rangle = \langle x, Ty \rangle$ for all $x \in X$, $y \in Y$. Show that $T \in \mathcal{B}(X, Y)$. Hint: Show that T is a closed operator.

(16) Give a direct proof for Proposition 2.11, without using Propositions 2.6 and 2.4.

(17) Let X and Y be normed linear spaces and $T : X \to Y$ be a linear operator. Show that T is bounded below if and only if

$$\gamma := \inf\{\|Tx\| : \|x\| = 1\} > 0,$$

and in that case, the norm of $T^{-1} : R(T) \to X$ is $1/\gamma$.

(18) Let X and Y be a normed linear space, $T \in \mathcal{B}(X, Y)$. Show that T is not bounded below if and only if there exists (x_n) in X_0 such that $\|x_n\| = 1$ for all $n \in \mathbb{N}$ and $\|Tx_n\| \to 0$ as $n \to \infty$.

(19) Let X be a Banach space and let $\mathcal{G}(X)$ be the set of all bijective operators in $\mathcal{B}(X)$. Show that, if $A \in \mathcal{G}(X)$, then

$$\{B \in \mathcal{B}(X) : \|A - B\| \leq 1/\|A^{-1}\|\} \subseteq \mathcal{G}(X).$$

(20) Let X be a Banach space. Show that $\{T \in \mathcal{B}(X) : 0 \in \sigma(T)\}$ is a closed subset of $\mathcal{B}(X)$.

(21) If $T \in \mathcal{K}(X)$ and X is infinite dimensional, then show that $0 \in \sigma_a(T)$.

(22) Let X be a Hilbert space and $T \in \mathcal{B}(X)$ be a normal operator. Show that $\|Tx\| = \|T^*x\|$ for all $x \in X$.

(23) For $u \in C[a, b]$, let $(Ax)(t) = u(t)x(t)$ for $x \in L^2[a, b]$ and $t \in [a, b]$. Show that A is a normal operator on $L^2[a, b]$. Taking X to be either $C[a, b]$ with $\|\cdot\|_\infty$ or $L^2[a, b]$, show that $\sigma(A) = [a, b]$. (*Hint:* Show that, for $\lambda \in \mathbb{K}$, $A - \lambda I$ is bounded below if and only if $\lambda \notin [a, b]$.)

(24) Justify: Let X and Y be Hilbert spaces and $T \in \mathcal{K}(X, Y)$. Then there exists a sequence (T_n) of finite rank operators in $\mathcal{B}(X, Y)$ such that $\|T - T_n\| \to 0$ as $n \to \infty$. (*Hint:* Use singular value representation.)

Chapter 3

Well-Posed Equations and Their Approximations

3.1 Introduction

In this chapter we consider the problem of approximately solving the operator equation

$$Tx = y, \qquad (3.1)$$

where $T : X \to Y$ is a bounded linear operator between normed linear spaces X and Y, and $y \in Y$.

We shall assume that (3.1) is **well-posed**, that is, T is bijective and $T^{-1} : Y \to X$ is continuous. Thus, for every $y \in Y$, there exists a unique $x \in X$ such that (3.1) is satisfied, and small perturbations in y lead only to small variations in the corresponding solution. Recall from Theorem 2.15 that if X and Y are Banach spaces, then continuity of T^{-1} is a consequence of the bijectivity of T.

By an **approximation method** for obtaining approximate solutions for (3.1) we mean a countable family

$$\mathcal{A} := \{(X_n, Y_n, T_n, y_n) : n \in \mathbb{N}\}$$

of quadruples, where, for each $n \in \mathbb{N}$, X_n and Y_n are subspaces of X and Y, respectively, and $T_n : X_n \to Y_n$ is a bounded linear operator and $y_n \in Y_n$. We say that such a method is **convergent** if there exists $N \in \mathbb{N}$ such that for every $n \geq N$, the equation

$$T_n x_n = y_n \qquad (3.2)$$

is uniquely solvable and (x_n) converges to a solution x of equation (3.1). Once we have a convergent method, one would also like to obtain estimates for the error $\|x - x_n\|$ for $n \geq N$ which would enable us to infer the *rate of convergence* of (x_n).

In applications, the space X_n may be finite dimensional which is close to X in some sense for large enough $n \in \mathbb{N}$. Then the problem of solving equation (3.2) is reduced to that of solving a system of linear equations involving scalar variables. Let us see how it happens.

Suppose X_n is a finite dimensional subspace with $\dim(X_n) = k_n$. Let $\{u_1, \ldots, u_{k_n}\}$ be a basis of X_n. Since every solution $x_n \in X_n$ of equation (3.2), if exists, is of the form

$$x_n = \sum_{j=1}^{k_n} \alpha_j u_j \tag{3.3}$$

for some scalars $\alpha_1, \ldots, \alpha_{k_n}$, the problem of solving (3.2) is equivalent to the problem of finding scalars α_j, $j = 1, \ldots, k_n$, such that

$$\sum_{j=1}^{k_n} \alpha_j T_n u_j = y_n.$$

Obviously, for the above problem to have a solution, it is necessary that y_n lie in $T_n(X_n)$. Without loss of generality, assume that $Y_n := T_n(X_n)$ and $y_n \in Y_n$. Let $\ell_n := \dim(Y_n)$, and let $\{v_1, \ldots, v_{\ell_n}\}$ be a basis of Y_n. Then there exist scalars $\beta_1, \ldots \beta_{\ell_n}$ such that $y_n = \sum_{i=1}^{\ell_n} \beta_i v_i$. Let a_{ij} be the $k_n \times \ell_n$ matrix of scalars such that

$$T_n u_j = \sum_{i=1}^{\ell_n} a_{ij} v_i, \quad j = 1, \ldots, k_n.$$

Thus, if x_n in (3.3) is a solution of equation (3.2), then $\alpha_1, \ldots, \alpha_{k_n}$ satisfy the system of equations

$$\sum_{j=1}^{k_n} a_{ij} \alpha_j = \beta_i, \quad i = 1, \ldots, \ell_n. \tag{3.4}$$

Conversely, if $\alpha_1, \ldots, \alpha_{k_n}$ are scalars such that (3.4) is satisfied, then it is seen that

$$x_n := \sum_{j=1}^{k_n} \alpha_j u_j$$

satisfies equation (3.2).

Exercise 3.1. Justify the last two statements.

We know that if X is infinite dimensional and T is a compact operator, then T does not have a continuous inverse (cf. Corollary 2.1). Consequently, for a compact operator T defined on an infinite dimensional normed linear space, the operator equation (3.1) is *ill-posed*.

We note that if X is a Banach space and T is of the form $T := \lambda I - A$ with $A \in \mathcal{B}(X)$ and λ is a known scalar, then equation (3.1),

$$\lambda x - Ax = y, \tag{3.5}$$

is well-posed if and only if $\lambda \notin \sigma(A)$.

Operator equation of the form (3.1) in which T does not have a continuous inverse is usually called an **equation of the first kind**, and a well-posed equation of the form (3.5) is called an **equation of the second kind**. Major part of this chapter is devoted to the study of operator equations of the second kind.

Recall that, for a compact operator A on a Banach space, the spectrum is a countable set, and 0 is the only possible limit point of the spectrum, so that outside every open neighbourhood of the origin of $\mathbb{K}$, there can be only a finite number of spectral values which are in fact eigenvalues (see Section 2.3.3). Thus, if λ is a nonzero scalar which is not an eigenvalue of a compact operator A on a Banach space, then the operator equation (3.5) is well-posed.

In order to obtain approximate solutions for a second kind operator equation (3.5), we may approximate A by a sequence (A_n) in $\mathcal{B}(X)$. Thus, in place of equation (3.2), we have

$$\lambda x_n - A_n x_n = y_n. \tag{3.6}$$

In applications, usually the operators A_n are of finite rank. But, $T_n = \lambda I - A_n$ is not of finite rank if the space X is infinite dimensional. In this special case, we now present an alternate way of obtaining approximate solutions as follows:

Suppose that $A_n \in \mathcal{B}(X)$ is of finite rank. Let $R(A_n)$ be spanned by $\{v_1, \ldots, v_{k_n}\}$. Then, we know (cf. Section 2.1.3) that there exist continuous linear functionals $f_1, \ldots, f_{k_n}$ on X such that

$$A_n x = \sum_{j=1}^{k_n} f_j(x) v_j \quad \forall x \in X.$$

Thus, a solution x_n of (3.6), if exists, satisfies the equation

$$x_n = \frac{1}{\lambda}(y_n + A_n x_n) = \frac{1}{\lambda}\left[y_n + \sum_{j=1}^{k_n} f_j(x_n) v_j \right].$$

On applying f_i to the above equation, we obtain

$$f_i(x_n) = \frac{1}{\lambda}\Big[f_i(y_n) + \sum_{j=1}^{k_n} f_j(x_n)f_i(v_j)\Big],$$

showing that the scalars $\alpha_j := f_j(x_n)$, $j = 1, \ldots, k_n$, satisfy the system of equations

$$\lambda\,\alpha_i - \sum_{j=1}^{k_n} \kappa_{ij}\alpha_j = \gamma_i, \quad i = 1, \ldots, k_n, \tag{3.7}$$

where

$$\kappa_{ij} = f_i(v_j), \quad \gamma_i = f_i(y_n)$$

for $i, j = 1, \ldots, k_n$. Conversely, if scalars $\alpha_1, \ldots, \alpha_{k_n}$ satisfy (3.7), then

$$x_n := \frac{1}{\lambda}\Big[y_n + \sum_{j=1}^{k_n} \alpha_j v_j\Big]$$

is a solution of (3.6). Indeed,

$$f_i(x_n) = \frac{1}{\lambda}\Big[f_i(y_n) + \sum_{j=1}^{k_n} \alpha_j f_i(v_j)\Big] = \frac{1}{\lambda}\Big[\gamma_i + \sum_{j=1}^{k_n} \kappa_{ij}\alpha_j\Big] = \alpha_i.$$

so that

$$x_n := \frac{1}{\lambda}\Big[y_n + \sum_{j=1}^{k_n} f_j(x_n)v_j\Big] = \frac{1}{\lambda}(y_n + A_n x_n).$$

NOTE: In the above discussion, the scalars α_j, β_j and vectors u_j, v_j would depend on n, and hence, strictly speaking, we should have written $\alpha_j^{(n)}$, $\beta_j^{(n)}$, $u_j^{(n)}$, $v_j^{(n)}$ in place of α_j, β_j, u_j, v_j. We avoided the superscripts to make the exposition less cumbersome.

3.2   Convergence and Error Estimates

Consider an approximation method

$$\mathcal{A} := \{(X_n, Y_n, T_n, y_n) : n \in \mathbb{N}\}$$

for (3.1). We assume that X and Y are Banach spaces, $T \in \mathcal{B}(X)$, and for each $n \in \mathbb{N}$, X_n and Y_n are closed subspaces of X and Y, respectively.

A natural way the sequence (y_n) in Y can be an approximation of $y \in Y$ is to have the convergence

$$\|y - y_n\| \to 0 \quad \text{as} \quad n \to \infty.$$

But the sequence (T_n) has to be an approximation of T in such a way that equation (3.2) is uniquely solvable for all large enough n and the sequence (x_n) converges to x as $n \to \infty$, where $Tx = y$. Also, we would like to get error estimates in terms of T_n and y_n. In this regard, we introduce the following definition:

A sequence (T_n) in $\mathcal{B}(X, Y)$ is said to be **stable** with a *stability index* or simply an *index* $N \in \mathbb{N}$, if

(a) $(\|T_n\|)$ is bounded,

(b) T_n is bijective for all $n \geq N$, and

(c) $\{\|T_n^{-1}\| : n \geq N\}$ is bounded.

For analogous definitions of stability of a sequence of operators, one may see Chatelin [10] and Prössdorf and Silbermann [64].

First we shall consider the case of $X_n = X$ and $Y_n = Y$ for all $n \in \mathbb{N}$. Thus, the method under consideration is

$$\mathcal{A}_0 := \{(X, Y, T_n, y_n) : n \in \mathbb{N}\}.$$

The following proposition essentially specifies conditions on (T_n) which ensures well-posedness of (3.1).

Proposition 3.1. *Suppose there exists $N \in \mathbb{N}$ such that T_n is invertible in $\mathcal{B}(X, Y)$ for all $n \geq N$ and $\|(T - T_n)x\| \to 0$ as $n \to \infty$ for every $x \in X$. Then the following hold:*

(i) *If $\{\|T_n^{-1}\|\}_{n \geq N}$ is bounded, then T is injective.*

(ii) *If $y \in Y$ is such that $(T_n^{-1}y)$ converges, then $y \in R(T)$.*

In particular, if (T_n) is stable, $\|(T - T_n)x\| \to 0$ as $n \to \infty$ for every $x \in X$ and $(T_n^{-1}y)$ converges for every $y \in Y$, then (3.1) is well-posed.

Proof. (i) Suppose $\{\|T_n^{-1}\|\}_{n \geq N}$ is bounded, say $\|T_n^{-1}\| \leq c$ for all $n \geq N$. Let $x \in X$ be such that $Tx = 0$. Then, for all $n \geq N$,

$$\|x\| = \|T_n^{-1}T_n x\| = \|T_n^{-1}(T_n - T)x\| \leq c\|(T_n - T)x\| \to 0$$

as $n \to \infty$. Hence, $x = 0$.

(ii) Since $\|(T - T_n)x\| \to 0$ for every $x \in X$, as a consequence of Uniform Boundedness Principle, there exists $b > 0$ such that $\|T_n\| \leq b$ for all $n \in \mathbb{N}$. Let $y \in Y$ be such that $(T_n^{-1}y)$ converges, say $x := \lim_{n \to \infty} T_n^{-1}y$. Note that

$$\|y - T_n x\| = \|T_n(T_n^{-1}y - x)\| \leq b\|T_n^{-1}y - x\| \to 0 \quad \text{as} \quad n \to \infty.$$

Hence,

$$\|y - Tx\| \le \|y - T_n x\| + \|(T_n - T)x\| \to 0 \quad \text{as} \quad n \to \infty.$$

Therefore, $y = Tx$.

The particular case is obvious. $\qquad\qquad\qquad\qquad\qquad\qquad\square$

We may observe that if x_n is a solution of (3.2), then for any $x \in X$,

$$T_n(x - x_n) = T_n x - y_n. \tag{3.8}$$

Now, we prove a theorem on convergence and error estimates.

Theorem 3.1. *Suppose (T_n) is stable with index $N \in \mathbb{N}$. Let x_n be the unique solution of (3.2) for $n \ge N$. Then for any $x \in X$,*

$$c_1 \|T_n x - y_n\| \le \|x - x_n\| \le c_2 \|T_n x - y_n\|, \tag{3.9}$$

where c_1 and c_2 are positive real numbers such that

$$\|T_n\| \le 1/c_1 \quad \text{and} \quad \|T_n^{-1}\| \le c_2 \quad \forall n \ge N.$$

If, in addition, $x \in X$ and $y \in Y$ are such that

$$\alpha_n := \|(T - T_n)x + (y_n - y)\| \to 0 \quad \text{as} \quad n \to \infty,$$

then

$$x_n \to x \iff Tx = y,$$

and in that case, $\alpha_n = \|T_n x - y_n\|$.

Proof.   Let $n \ge N$. Then, from (3.8), we have

$$\|x - x_n\| = \|T_n^{-1}(T_n x - y_n)\| \le \|T_n^{-1}\| \, \|T_n x - y_n\|,$$

$$\|T_n x - y_n\| = \|T_n(x - x_n)\| \le \|T_n\| \, \|x - x_n\|$$

so that

$$\frac{\|T_n x - y_n\|}{\|T_n\|} \le \|x - x_n\| \le \|T_n^{-1}\| \, \|T_n x - y_n\|.$$

From this, we obtain (3.9). The inequalities in (3.9) imply that

$$x_n \to x \iff \|T_n x - y_n\| \to 0.$$

Further, we observe that

$$\alpha_n \to 0 \iff T_n x - y_n \to Tx - y.$$

The assumption $\alpha_n \to 0$ together with the above two equivalence imply that $x_n \to x$ if and only if $Tx = y$, and in that case, $\alpha_n = \|T_n x - y_n\|$. $\quad\square$

A particular case of Theorem 3.1 is worth stating as a corollary.

Corollary 3.1. *Suppose T is bijective and (T_n) is stable with index $N \in \mathbb{N}$. Let x be the solution of (3.1) and x_n be the solution of (3.2) for $n \geq N$. Suppose further that*

$$y_n \to y \quad \text{and} \quad (T - T_n)x \to 0 \quad \text{as} \quad n \to \infty.$$

Then $\alpha_n := \|T_n x - y_n\| \to 0$ as $n \to \infty$ and

$$c_1 \alpha_n \leq \|x - x_n\| \leq c_2 \alpha_n \quad \forall n \geq N,$$

where c_1 and c_2 are positive real numbers such that $\|T_n\| \leq 1/c_1$ and $\|T_n^{-1}\| \leq c_2$ for all $n \geq N$. In particular, $x_n \to x$ as $n \to \infty$.

Exercise 3.2. Write details of the proof of Corollary 3.1.

Remark 3.1. In applications, x_n may be obtained as a result of computation and x may be unknown. Hence, in Theorem 3.1 and in Corollary 3.1, the quantity α_n is of *a priori* nature as it involves the 'unknown' x. If T is invertible, then from the relation

$$T(x_n - x) = Tx_n - y,$$

we obtain

$$\frac{1}{\|T\|} \beta_n \leq \|x - x_n\| \leq \|T^{-1}\| \beta_n, \tag{3.10}$$

where $\beta_n := \|Tx_n - y\|$. Note that the quantity β_n can be thought of as computational as it is in terms of known quantities. Thus, combining (3.9) and (3.10), we have

$$\kappa_1 \max\{\alpha_n, \beta_n\} \leq \|x - x_n\| \leq \kappa_2 \min\{\alpha_n, \beta_n\}$$

where

$$\kappa_1 = \min\{c_1, \|T\|^{-1}\} \quad \text{and} \quad \kappa_2 = \max\{c_2, \|T^{-1}\|\}.$$

Since

$$\alpha_n := \|T_n x - y_n\| \leq \|(T - T_n)x\| + \|y_n - y\|,$$
$$\beta_n := \|Tx_n - y\| \leq \|(T - T_n)x_n\| + \|y_n - y\|,$$

we have

$$\|x - x_n\| \leq \kappa_2 \varepsilon_n,$$

where

$$\varepsilon_n := \|y - y_n\| + \min\{\|(T - T_n)x\|, \|(T - T_n)x_n\|\}.$$

Thus, the quality of the approximation (x_n) can be inferred from the quality of the approximations (y_n) and (T_n). We shall elaborate this point at a later context.   ◇

NOTATION: For sequences (a_n) and (b_n) of positive reals such that $a_n \to 0$ and $b_n \to 0$ as $n \to \infty$, we shall use the notation

$$a_n = O(b_n)$$

to mean that there exist $\kappa > 0$ and $N \in \mathbb{N}$ such that

$$a_n \leq \kappa \, b_n \quad \forall \, n \geq N.$$

If $a_n = O(b_n)$ and $b_n = O(a_n)$, then we shall write

$$a_n \approx b_n.$$

We shall use c_1, c_2, etc., for generic positive constants which may take different values at different contexts, but independent of n.

Now, we consider the case, where the spaces X_n and Y_n need not be equal to X and Y, respectively. So, for each $n \in \mathbb{N}$, let $T_n \in \mathcal{B}(X_n, Y_n)$ and $y_n \in Y_n$. We would like to have sufficient conditions under which (3.2) is uniquely solvable for all large enough n, say for $n \geq N$ and $x_n \to x$ as $n \to \infty$, where x is the unique solution of (3.1). One would also like to obtain estimates for the error $\|x - x_n\|$ for $n \geq N$. In order to do this, we first assume that there exists a sequence (P_n) of projections in $\mathcal{B}(X)$ such that $R(P_n) = X_n$, $n \in \mathbb{N}$. We may recall from Sections 2.2.3 and 2.2.4 that existence of a continuous projection onto X_n is guaranteed if either X_n is finite dimensional or if X is a Hilbert space and X_n is a closed subspace. A specific example of such a projection when $X = C[a, b]$ is considered in Section 3.4.1.

Theorem 3.2. *Assume that (T_n) is stable with index N and let x_n be the solution of (3.2) for all $n \geq N$. Then for any $x \in X$,*

$$\|x - x_n\| \leq \|x - P_n x\| + c\|T_n P_n x - y_n\|, \tag{3.11}$$

where $c \geq \|T_n^{-1}\|$ for all $n \geq N$. Further, if $x \in X$ and $y \in Y$ are such that

$$\tilde{\alpha}_n := \|(T_n P_n - T)x + (y - y_n)\| \to 0$$

and $P_n x \to x$ as $n \to \infty$, then

$$x_n \to x \iff Tx = y,$$

and in that case, $\tilde{\alpha}_n = \|T_n P_n x - y_n\|$.

Proof.   For $x \in X$, we have

$$x - x_n = (x - P_n x) + T_n^{-1}(T_n P_n x - y_n). \tag{3.12}$$

From this, the estimate in (3.11) follows.

Next assume that $\tilde{\alpha}_n := \|(T_n P_n - T)x + (y - y_n)\| \to 0$ and $P_n x \to x$ as $n \to \infty$ for some $x \in X$, $y \in Y$. Then

$$T_n P_n x - y_n \to Tx - y \quad \text{as} \quad n \to \infty.$$

Therefore, (3.12) together with boundedness of the sequences $(\|T_n\|)$ and $(\|T_n^{-1}\|)_{n \geq N}$ imply that $x_n \to x$ if and only if $Tx = y$, and in that case, $\tilde{\alpha}_n = \|T_n P_n x - y_n\|$. $\square$

Now, analogous to Corollary 3.1 we have the following result.

Corollary 3.2. *Suppose T is bijective and (T_n) is stable with index N. Let x be the solution of (3.1) and x_n be the solution of (3.2) for all $n \geq N$. Suppose further that*

$$y_n \to y, \quad P_n x \to x, \quad (T_n P_n - T)x \to 0$$

as $n \to \infty$. Then

$$\|x - x_n\| \leq \|x - P_n x\| + c\|T_n P_n x - y_n\|,$$

where $c > 0$ is such that $\|T_n^{-1}\| \leq c$ for all $n \geq N$. In particular,

$$x_n \to x \quad \text{as} \quad n \to \infty.$$

Exercise 3.3. Write details of the proof of Corollary 3.2.

3.3 Conditions for Stability

In this section we discuss some sufficient conditions on the approximation properties of the sequence (T_n) of operators which ensure its stability when $X = X$ and $Y_n = Y$. A simplest assumption that one may impose on (T_n) would be to have **pointwise convergence**, that is, for each $x \in X$,

$$\|Tx - T_n x\| \to 0 \quad \text{as} \quad n \to \infty,$$

and in that case, we also say that (T_n) is a **pointwise approximation** of T, and write $T_n \to T$ **pointwise**.

We may observe that, if (T_n) is a pointwise approximation of T, then by Banach Steinhaus theorem (Theorem 2.11), the sequence $(\|T_n\|)$ is bounded.

Example 3.1. Suppose $T \in \mathcal{B}(X,Y)$, and (P_n) and (Q_n) are projection operators in $\mathcal{B}(X)$ and $\mathcal{B}(Y)$, such that for each $x \in X$ and $y \in R(T)$,

$$P_n x \to x \quad \text{and} \quad Q_n y \to y \quad \text{as} \quad n \to \infty.$$

Then we have

$$\|TP_nx - Tx\| = \|T(P_nx - x)\| \leq \|T\|\,\|P_nx - x\| \to 0,$$

and

$$\|Q_nTx - Tx\| \to 0$$

as $n \to \infty$ for every $x \in X$. Thus, (TP_n) and (Q_nT) are pointwise approximations of T.

Suppose, in addition, $(\|Q_n\|)$ is bounded. Then we have

$$
\begin{aligned}
\|Q_nTP_nx - Tx\| &\leq \|Q_n(TP_nx - Tx)\| + \|Q_nTx - Tx\| \\
&\leq \|Q_n\|\,\|TP_nx - Tx\| + \|Q_nTx - Tx\| \\
&\to 0
\end{aligned}
$$

as $n \to \infty$ for every $x \in X$. Thus, in this case, (Q_nTP_n) is also a pointwise approximation of T.   ◇

Exercise 3.4. Let X be a separable Hilbert space and $\{u_n : n \in \mathbb{N}\}$ be an orthonormal basis of X. For $n \in \mathbb{N}$, let

$$P_nx = \sum_{i=1}^{n} \langle x, u_i \rangle u_i, \quad x \in X.$$

Prove that
  (i) P_n is an orthogonal projection for each $n \in \mathbb{N}$, and
  (ii) $\|P_nx - x\| \to 0$ as $n \to \infty$ for each $x \in X$.

Pointwise approximation is, in fact, too weak to satisfy even well-posedness of the approximating equation. For example, suppose (P_n) is a sequence of projection operators in $\mathcal{B}(X)$ such that $P_nx \to x$ as $n \to \infty$ for each $x \in X$. Then it is obvious that equation (3.2) is not well-posed with $T_n = P_n$, unless $P_n = I$.

So we are looking for stronger approximation properties for (T_n) which guarantee solvability of (3.2) and convergence of (x_n) to x.

3.3.1 *Norm approximation*

A sequence (T_n) in $\mathcal{B}(X,Y)$ is called a **norm approximation** of T if

$$\|T - T_n\| \to 0 \quad \text{as} \quad n \to \infty.$$

If (T_n) is a norm approximation of T, then we may also say that (T_n) **converges to T in norm**, and write $T_n \to T$ **in norm**.

Remark 3.2. We may observe that if $T : X \to X$ is of the form $\lambda I - A$ for some nonzero scalar, and if (A_n) in $\mathcal{B}(X)$ is a norm approximation of $A \in \mathcal{B}(X)$, then taking $T_n : \lambda I - A_n$, we have

$$\|T - T_n\| = \|A - A_n\| \to 0$$

as $n \to \infty$. $\Diamond$

Theorem 3.3. *Suppose T is bijective and (T_n) is a norm approximation of T. Then (T_n) is stable.*

Proof. For $n \in \mathbb{N}$, let $T_n = T + E_n$, where $E_n = T_n - T$. By hypothesis $\|E_n T^{-1}\| \to 0$ as $n \to \infty$. Let $N \in \mathbb{N}$ be such that $\|E_n T^{-1}\| \le 1/2$ for all $n \ge N$. Then, by Corollary 2.13, T_n is bijective and

$$\|T_n^{-1}\| \le \frac{\|T^{-1}\|}{1 - \|E_n T^{-1}\|} \le 2\|T^{-1}\| \quad . \tag{3.13}$$

for all $n \ge N$. Thus, (T_n) is stable. $\square$

Let us state explicitly the implication of the above theorem for the operator equation (3.1) which follows from Corollary 3.2.

Corollary 3.3. *Suppose T is bijective and (T_n) is a norm approximation of T. Then there exists $N \in \mathbb{N}$ such that (3.2) is uniquely solvable for all $n \ge N$, and*

$$\|x - x_n\| \le c\|(T - T_n)x\| + \|y - y_n\| \quad \forall n \ge N,$$

where c is a positive number and x and x_n are the solutions of (3.1) and (3.2) respectively. In particular, if $y_n \to y$ as $n \to \infty$, then

$$x_n \to x \quad as \quad n \to \infty.$$

Example 3.2. Suppose $P_n \in \mathcal{B}(X)$ and $Q_n \in \mathcal{B}(Y)$ are projection operators such that $P_n x \to x$ and $Q_n y \to y$ as $n \to \infty$ for each $x \in X$ and $y \in Y$. Let $K : X \to X$ be a compact operator. Then by Theorem 2.13,

$$\|Q_n K - K\| \to 0 \quad as \quad n \to \infty.$$

Thus, $(Q_n K)$ is a norm approximations of K.

Now, suppose X and Y are Hilbert spaces, and (P_n) is a sequence of orthogonal projections. Then by compactness of K^*, the adjoint of K, we have

$$\|K P_n - K\| = \|P_n K^* - K^*\| \to 0.$$

Hence, by the boundedness of $(\|Q_n\|)$ (cf. Theorem 2.11), we also have

$$\|Q_n K P_n - K\| \leq \|Q_n(K P_n - K)\| + \|Q_n K - K\|$$
$$\leq \|Q_n\| \, \|K P_n - K\| + \|Q_n K - K\|$$
$$\to 0$$

as $n \to \infty$. Thus, in this case, $(K P_n)$ and $(Q_n K P_n)$ are also norm approximations of K.

Next, suppose that $X = Y$,

$$T = \lambda I - K \quad \text{and} \quad T_n : \lambda I - K_n,$$

where λ is a nonzero scalar and K_n belongs to $\{Q_n K, K P_n, Q_n K P_n\}$. By Remark 3.2, it follows that (T_n) is a norm approximation of T if $K_n = Q_n K$ or if X is a Hilbert space, P_n are orthogonal projections and K_n belongs to $\{K P_n, Q_n K P_n\}$. ◊

Remark 3.3. (a) Recall that if P is a projection, then $I - P$ is also a projection. Hence, $\|I - P\| \geq 1$ whenever $P \neq I$. Thus, if (P_n) is a sequence of projections such that $P_n \neq I$ for infinitely many n, then (P_n) cannot be a norm approximation of I.

(b) Suppose $T \in \mathcal{B}(X, Y)$ is invertible, and (P_n) and (Q_n) are sequences of (non-identity) projections in $\mathcal{B}(X)$ and $\mathcal{B}(Y)$ respectively. Then neither $(Q_n T)$ nor $(T P_n)$ is a norm approximation of T. ◊

Exercise 3.5. Justify the last statement in Remark 3.3(b).

Example 3.3. Consider the integral operator K defined by

$$(Kx)(s) = \int_a^b k(s, t) x(t) dt, \quad a \leq s \leq b, \quad x \in C[a, b], \qquad (3.14)$$

where $k(\cdot, \cdot) \in C([a, b] \times [a, b])$. Then we know (See Examples 2.9 and 2.16) that $K : C[a, b] \to C[a, b]$ is a compact operator on $C[a, b]$ with $\| \cdot \|_\infty$, and

$$\|K\| \leq \sup_{a \leq s \leq b} \int_a^b |k(s, t)| dt.$$

For each $n \in \mathbb{N}$, let $k_n(\cdot, \cdot)$ be a continuous function such that

$$\sup_{a \leq s \leq b} \int_a^b |k(s, t) - k_n(s, t)| dt \to 0$$

as $n \to \infty$. Note that the above condition of convergence is satisfied if, for example, $k_n(\cdot, \cdot)$ is such that

$$\sup_{s,t \in [a,b]} |k(s,t) - k_n(s,t)| \to 0 \quad \text{as} \quad n \to \infty.$$

We know that *Weierstrass approximation theorem* guarantees the existence of such a sequence $(k_n(\cdot, \cdot))$ for every continuous $k(\cdot, \cdot)$. Now, we define $K_n : C[a,b] \to C[a,b]$ by

$$(K_n x)(s) = \int_a^b k_n(s,t)x(t)dt, \quad a \le s \le b, \quad x \in C[a,b].$$

Then it follows that

$$\|K - K_n\| \le \sup_{a \le s \le b} \int_a^b |k(s,t) - k_n(s,t)|dt \to 0$$

as $n \to \infty$. Thus, (K_n) is a norm approximation of K. Hence, for any nonzero scalar, if we take

$$T := \lambda I - K \quad \text{and} \quad T_n := \lambda I - K_n$$

then (T_n) is also a norm approximation of T. ◇

Example 3.4. Let $K : X \to Y$ be a compact operator of infinite rank, where X and Y are Hilbert spaces. Consider the singular value representation

$$Kx = \sum_{j=1}^{\infty} \sigma_j \langle x, u_j \rangle v_j, \quad x \in X$$

of T (see Theorem 2.28). Recall that $\{u_n : n \in \mathbb{N}\}$ and $\{v_n : n \in \mathbb{N}\}$ are orthonormal bases of $N(K)^{\perp}$ and $\mathrm{cl}R(K)$ respectively, and (σ_n) is a sequence of positive real numbers such that $\sigma_n \to 0$ as $n \to \infty$. Now, let

$$K_n x = \sum_{j=1}^{n} \sigma_j \langle x, u_j \rangle v_j, \quad x \in X, \ n \in \mathbb{N}.$$

Then, for every $x \in X$ and for every $n \in \mathbb{N}$, we have

$$\|(K - K_n)x\|^2 = \sum_{j=n+1}^{\infty} \sigma_j^2 |\langle x, u_j \rangle|^2$$

$$\le (\max_{j>n} \sigma_j^2) \|x\|^2.$$

Thus,

$$\|K - K_n\| \le \max_{j>n} \sigma_j \to 0 \quad \text{as} \quad n \to \infty.$$

Thus, (K_n) is a norm approximation of K. ◇

We shall see that, in certain standard numerical approximation procedures, norm convergence is too strong to hold. In order to treat such cases, in the next subsection, we shall consider a requirement on (T_n) which is weaker than norm convergence.

3.3.2   *Norm-square approximation*

Suppose $T \in \mathcal{B}(X,Y)$ is bijective. Then a sequence (T_n) in $\mathcal{B}(X,Y)$ is called **norm-square approximation** of T if

$$\|[(T - T_n)T^{-1}]^2\| \to 0 \quad \text{as} \quad n \to \infty. \tag{3.15}$$

If (T_n) is a norm-square approximation of T, we also say that (T_n) converges to T in **norm-square sense**.

Clearly, norm convergence implies norm-square convergence. But, the converse need not hold. The following two simple examples illustrate this.

Example 3.5. Let $X = \mathbb{K}^2$ with any norm, and let

$$A = \begin{bmatrix} 1 & 0 \\ 0 & 1 \end{bmatrix}, \quad B = \begin{bmatrix} 1 & 1 \\ 0 & 1 \end{bmatrix}, \quad C = \begin{bmatrix} 0 & -1 \\ 0 & 0 \end{bmatrix}.$$

Then we have

$$A - B = C \quad \text{and} \quad [(A - B)A^{-1}]^2 = C^2 = 0.$$

Now, for $n \in \mathbb{N}$, let T, T_n, E be the operators corresponding to the matrices A, B, C and respectively. Then we have

$$T - T_n = E \quad \text{and} \quad [(T - T_n)T^{-1}]^2 = E^2 = 0$$

for all $n \in \mathbb{N}$. Thus, (T_n) is a norm-square approximation of T which is not a norm approximation.   ◊

Example 3.6. Let $X = Y = \ell^2$, $T = I$ and $T_n := I - A_n$, where A_n is defined by

$$A_n x = x(2)e_1 + \frac{x(n)}{n}e_n, \quad x \in \ell^2.$$

Then we see

$$\|T - T_n\| = \|A_n\| = 1 \quad \forall n \in \mathbb{N},$$

whereas

$$\|[(T - T_n)T^{-1}]^2\| = \|A_n^2\| = \frac{1}{n^2} \to 0 \quad \text{as} \quad n \to \infty.$$

Thus, (T_n) is a norm-square approximation of T which is not norm approximation.   ◊

In due course, we shall give examples of practical importance where we do not have norm convergence, but have norm-square convergence.

Theorem 3.4. *Suppose T is bijective, (T_n) is a norm-square approximation of T and $(\|T_n\|)$ is bounded. Then (T_n) is stable.*

Proof. For $n \in \mathbb{N}$, let $T_n = T + E_n$, where $E_n = T_n - T$. Since $(\|T_n\|)$ is bounded, there exists $c > 0$ such that $\|E_n T^{-1}\| \leq c$ for all $n \in \mathbb{N}$. Also, the hypothesis $\| (E_n T^{-1})^2 \| \to 0$ implies that there exists a positive integer N such that

$$\| (E_n T^{-1})^2 \| \leq 1/2 \quad \forall n \geq N.$$

Therefore, by Theorem 2.22, the operator T_n is bijective and

$$\|T_n^{-1}\| \leq \frac{\|T^{-1}\| \left(1 + \|E_n T^{-1}\|\right)}{1 - \| (E_n T^{-1})^2 \|} \leq 2(1 + c)\|T^{-1}\| \tag{3.16}$$

for all $n \geq N$. Thus, (T_n) is stable. $\qquad\square$

Exercise 3.6. Give one example each for the following:

(i) Norm-square approximation does not imply pointwise approximation.

(ii) Pointwise approximation does not imply norm-square approximation.

3.4 Projection Based Approximation

We have seen in Remark 3.3(b) that if T is invertible in $\mathcal{B}(X, Y)$, and if (P_n) and (Q_n) are sequences of non-identity projection operators in $\mathcal{B}(X)$ and $\mathcal{B}(Y)$ respectively, then neither $(Q_n T)$ nor $(T P_n)$ can be a norm approximation of T. Thus, in order that $(Q_n T)$ or $(T P_n)$ to be a norm approximation, it is necessary that T is not invertible in $\mathcal{B}(X, Y)$. In fact, we have seen in Example 3.2 that if Y is a Banach space, (Q_n) is a sequence of projections in $\mathcal{B}(Y)$ such that $Q_n \to I$ pointwise and $K \in \mathcal{K}(X, Y)$, then $(Q_n K)$ is a norm approximations of K. We have also seen in Example 3.2 that if, in addition, X and Y are Hilbert spaces, and (P_n) is a sequence of orthogonal projections on X such that $P_n \to I$ pointwise, then both $(K P_n)$ and $(Q_n K P_n)$ are norm approximations of $K \in \mathcal{K}(X, Y)$.

In the following subsection, we shall discuss a special class of projections on the space $C[a, b]$ which will be used in later sections.

3.4.1 *Interpolatory projections*

Let $t_1, \ldots, t_N$ be distinct points in $[a, b]$ and let X be the space $C[a, b]$ or $B[a, b]$ (the space of all bounded functions on $[a, b]$). Let $\{u_1, \ldots, u_N\}$ be a set of functions in X such that

$$u_i(t_j) = \begin{cases} 1 \text{ if } i = j \\ 0 \text{ if } i \neq j \end{cases} \tag{3.17}$$

Then the operator $P : X \to X$ defined by

$$Px = \sum_{i=1}^{N} x(t_i)u_i, \quad x \in X,$$

is a linear operator with $R(P) = \text{span}\{u_1, \ldots, u_N\}$. We note that, for every $x \in X$,

$$(Px)(t_j) = x(t_j) \quad \forall j \in \{1, \ldots, N\}$$

so that P is a projection operator. This projection operator is called an **interpolatory projection** associated with (t_i, u_i), $i = 1, \ldots, N$.

Exercise 3.7. Show that the functions $u_1, \ldots, u_N$ considered above satisfying (3.17) are linearly independent, and the projection P is a bounded linear operator on $C[a, b]$ with respect to the norm $\|\cdot\|_\infty$ and $\|P\| = \sum_{i=1}^{N} \|u_i\|_\infty$.

A particular case of an interpolatory projection is the so-called **Lagrange interpolatory projection** in which

$$u_i(t) = \ell_i(t) := \prod_{j \neq i} \frac{t - t_j}{t_i - t_j}, \quad t \in [a, b].$$

Note that each $\ell_i(t)$ above is a polynomial of degree $N - 1$. For $x \in C[a, b]$, the polynomial

$$(Px)(t) := \sum_{i=1}^{N} x(t_i)\ell_i(t), \quad t \in [a, b],$$

is called the **Lagrange interpolatory polynomial** based on the nodes $t_1, \ldots, t_N$, which is the unique polynomial having the property $(Px)(t_j) = x(t_j)$, $j = 1, \ldots, N$.

For $n \in \mathbb{N}$, let $t_i^{(n)} \in [a, b]$ for $i = 1, \ldots, n$ be such that

$$a \leq t_0^{(n)} < t_1^{(n)} < \ldots < t_n^{(n)} \leq b$$

and

$$\max\{t_j^{(n)} - t_{j-1}^{(n)} : j = 1, \ldots, n\} \to 0, \quad \text{as} \quad n \to \infty.$$

Let P_n be the Lagrange interpolatory projection based on the nodes $t_i^{(n)}$, $i = 1, \ldots, n$. One may ask if we have the convergence $\|P_n x - x\|_\infty \to 0$ as

$n \to \infty$ for every $x \in C[a, b]$. Unfortunately, this need not hold (cf. [6]; see also, [51], Section 6.2).

Because of the above non-convergence result, it is useful to take piecewise polynomial interpolatory projections, that is, an interpolatory projection P_n such that for each $x \in C[a, b]$, $P_n x|_{(t_{i-1}^{(n)}, t_i^{(n)})}$ is a polynomial of degree k, for $i = 1, \ldots, n$, where

$$a = t_0^{(n)} \le t_1^{(n)} < t_2^{(n)} < \ldots \le t_n^{(n)} \le b$$

is a partition of $[a, b]$. A simple example of such an interpolatory projection is obtained by taking $k = 1$ as follows: For $i = 2, \ldots, n$ and $t \in [t_{i-1}^{(n)}, t_i^{(n)}]$, take the graph of $P_n x$ to be the straight line segment joining the points $(t_{i-1}^{(n)}, x(t_{i-1}^{(n)}))$ and $(t_i^{(n)}, x(t_i^{(n)}))$, and on $[a, t_1^{(n)}]$ and $[t_n^{(n)}, b]$, define $P_n x$ to be the constants $x(t_1^{(n)})$ and $x(t_n^{(n)})$, respectively. It can be seen that,

$$P_n x = \sum_{i=1}^{n} x(t_i^{(n)}) u_i^{(n)}, \quad x \in C[a, b], \tag{3.18}$$

where $u_i^{(n)}$ are the *hat functions* based on the nodes $t_1^{(n)}, \ldots, t_n^{(n)}$, that is,

$$u_1^{(n)}(t) = \begin{cases} 1 & \text{if } a \le t < t_1^{(n)} \\ \dfrac{t_2^{(n)} - t}{t_2^{(n)} - t_1^{(n)}} & \text{if } t_1^{(n)} \le t < t_2^{(n)} \\ 0 & \text{if } t_2^{(n)} \le t \le b, \end{cases}$$

$$u_n^{(n)}(t) = \begin{cases} 0 & \text{if } a \le t < t_{n-1}^{(n)} \\ \dfrac{t_n^{(n)} - t}{t_n^{(n)} - t_{n-1}^{(n)}} & \text{if } t_{n-1}^{(n)} \le t < t_n^{(n)} \\ 1 & \text{if } t_n^{(n)} \le t \le b \end{cases}$$

and for $i = 2, \ldots, n-1$,

$$u_i^{(n)}(t) = \begin{cases} 0 & \text{if } a \le t < t_{i-1}^{(n)} \\ \dfrac{t_{i-1}^{(n)} - t)}{t_{i-1}^{(n)} - t_i^{(n)}} & \text{if } t_{i-1}^{(n)} \le t < t_i^{(n)} \\ \dfrac{t_{i+1}^{(n)} - t}{t_{i+1}^{(n)} - t_i^{(n)}} & \text{if } t_i^{(n)} \le t < t_{i+1}^{(n)} \\ 0 & \text{if } t_{i+1}^{(n)} \le t \le b \end{cases}$$

It can be shown that the sequence (P_n) of interpolatory projections defined in (3.18) converges pointwise to the identity operator on $C[a, b]$ (see Problem 8).

For more examples of projections P_n such that $\|P_n x - x\| \to 0$ as $n \to \infty$, we refer the reader to Problem 9 and Limaye [38].

3.4.2 *A projection based approximation*

For $n \in \mathbb{N}$, let P_n be an interpolatory projection on $[a, b]$ based on the nodes $t_1^{(n)}, \ldots, t_n^{(n)}$ and functions $u_1^{(n)}, \ldots, u_n^{(n)}$ in $C[a, b]$ satisfying the condition (3.17), that is,

$$P_n x = \sum_{j=1}^{n} x(t_j^{(n)}) u_j^{(n)}, \quad x \in C[a, b]. \tag{3.19}$$

Let K be the integral operator in (3.14) with continuous kernel. Then, we see that $P_n K$ is an integral operator,

$$(P_n K x)(s) = \int_a^b k_n(s, t) x(t) \, dt, \quad x \in C[a, b], \; s \in [a, b],$$

with the kernel $k_n(s, t)$ is defined by

$$k_n(s, t) = \sum_{j=1}^{n} k(t_j^{(n)}, t) u_j^{(n)}(s), \quad s, t \in [a, b].$$

Note that the kernel $k_n(s, t)$ is a finite sum of products of functions of s and t. Such a kernel is called a *degenerate kernel* on $[a, b]$. More precisely, a function $\kappa(\cdot, \cdot)$ on $[a, b] \times [a, b]$ is said to be a **degenerate kernel** if it can be written as

$$\kappa(s, t) = \sum_{j=1}^{m} \phi_j(s) \psi_j(t), \quad s, t \in [a, b],$$

for some functions $\phi_1, \ldots, \phi_m, \psi_1, \ldots, \psi_m$ defined on $[a, b]$.

In case of $P_n \to I$ pointwise on $C[a, b]$ with respect to $\| \cdot \|_\infty$, then we know by Theorem 2.13 that

$$\|K - P_n K\| \to 0 \quad \text{as} \quad n \to \infty.$$

Exercise 3.8. In the above, if $\|x - P_n x\|_\infty \to 0$ as $n \to \infty$ for every $x \in C[a, b]$, then show that

$$\sup_{a \le s \le b} \int_a^b |k(s, t) - k_n(s, t)| dt \to 0 \quad \text{as} \quad n \to \infty.$$

Hint: See Example 3.3.

Now, we show that $(K P_n)$ is not a norm approximation of K. First we recall that K is a compact operator on $C[a, b]$ with $\| \cdot \|_\infty$, and

$$\|K\| = \sup_{a \le s \le b} \int_a^b |k(s, t)| dt.$$

Thus, using the continuity of the map $s \mapsto \int_a^b |k(s,t)|dt$, $s \in [a,b]$, there exists $\tau \in \Omega$ such that

$$\|K\| = \int_a^b |k(\tau,t)|dt.$$

The following lemma will be used in due course.

Lemma 3.1. *Let K be the integral operator in (3.14) with continuous kernel $k(\cdot,\cdot)$. Let $\tau \in [a,b]$ be such that $\int_\Omega |k(\tau,t)|dt = \|K\|$, and for $\varepsilon > 0$, let $\Omega_\varepsilon \subset [a,b]$ and $x_\varepsilon \in C[a,b]$ be such that*

$$\|x_\varepsilon\|_\infty \leq 1, \quad m(\Omega_\varepsilon) < \varepsilon,$$

and for $t \notin \Omega_\varepsilon$,

$$x_\varepsilon(t) = \begin{cases} |k(\tau,t)|/k(\tau,t) & \text{if } k(\tau,t) \neq 0 \\ 0 & \text{if } k(\tau,t) = 0, \end{cases}$$

where $m(\cdot)$ denotes the Lebesgue measure. Then

$$\|Kx_\varepsilon\|_\infty \to \|K\| \quad \text{as} \quad \varepsilon \to 0.$$

Proof. Taking $\Omega := [a,b]$ we have

$$|(Kx_\epsilon)(\tau)| = |\int_\Omega k(\tau,t)x_\epsilon(t)dt|$$

$$\geq \int_{\Omega - \Omega_\epsilon} |k(\tau,t)|dt - |\int_{\Omega_\epsilon} k(\tau,t)x_\epsilon(t)dt|$$

$$\geq \int_{\Omega - \Omega_\epsilon} |k(\tau,t)|dt - \epsilon \max_{s,t \in \Omega} |k(s,t)|.$$

In particular,

$$\|K\| \geq \|Kx_\varepsilon\|_\infty \geq |(Kx_\epsilon)(\tau)| \geq \int_{\Omega - \Omega_\epsilon} |k(\tau,t)|dt - \epsilon \max_{s,t \in \Omega} |k(s,t)|.$$

Using a continuous version of *monotone convergence theorem* (cf. [51]), it follows that

$$\int_{\Omega - \Omega_\epsilon} |k(\tau,t)|dt \to \int_\Omega |k(\tau,t)|dt = \|K\| \quad \text{as} \quad \epsilon \to 0.$$

Hence,

$$\|K\| \geq \lim_{\varepsilon \to 0} \|Kx_\varepsilon\|_\infty \geq \|K\|,$$

proving the required result. $\square$

Remark 3.4. It can be easily seen that for each $\varepsilon > 0$, set $\Omega_\varepsilon \subset [a,b]$ and $x_\varepsilon \in C[a,b]$ satisfying the requirements in Lemma 3.1 do exist. In the situations where we shall make use of the above lemma, we have $t_i^{(n)} \in [a,b]$, $i = 1, \ldots, n$, for each $n \in \mathbb{N}$, and Ω_ε and x_ε are to be chosen in such a way that

$$t_i^{(n)} \in \Omega_\varepsilon \quad \text{and} \quad x_\varepsilon(t_i^{(n)}) = 0, \quad i = 1, \ldots, n.$$

It can be seen that $\Omega_\varepsilon \subseteq [a,b]$ and $x_\varepsilon \in C[a,b]$ with the above additional requirements also exist. ◊

For $\varepsilon > 0$, let Ω_ε and $x_\varepsilon \in C[a,b]$ be as in Lemma 3.1 with additional conditions (see Remark 3.4)

$$t_i^{(n)} \in \Omega_\varepsilon \quad \text{and} \quad x_\varepsilon(t_i^{(n)}) = 0, \quad i = 1, \ldots, n.$$

Let P_n be as in (3.19). Then we see that $P_n x_\epsilon = 0$ so that

$$\|K - KP_n\| \geq \|(K - KP_n)x_\epsilon\|_\infty = \|Kx_\epsilon\|_\infty.$$

Letting $\varepsilon \to 0$, by Lemma 3.1, we have

$$\|K - KP_n\| \geq \|K\| \quad \forall n \in \mathbb{N}.$$

In particular, $\|K - KP_n\| \nrightarrow 0$ as $n \to \infty$.

3.5   Quadrature Based Approximation

A well considered numerical procedure for obtaining approximations for integral equations of the second kind,

$$\lambda x(s) - \int_a^b k(s,t)x(t)dt = y(s), \quad a \leq s \leq b,$$

is the so-called *Nyström method*. We assume that $k(\cdot, \cdot)$ is continuous on $[a,b] \times [a,b]$. The above equation can be written as

$$\lambda x - Kx = y,$$

where K is the integral operator as in (3.14) and $\lambda \neq 0$ is a scalar which is not an eigenvalue of K. In *Nyström method* that we discuss in Section 3.5.3, the integral appearing in the definition of K is approximated by a convergent quadrature rule giving an approximation (K_n) of K, called the *Nyström approximation* of K.

In the following two subsections, we describe a quadrature rule and the associated Nyström approximation, and show that (K_n) is a pointwise approximation of K, but not a norm approximation.

3.5.1 Quadrature rule

A **quadrature rule** for approximating the integral

$$\varphi(x) := \int_a^b x(t)dt, \quad x \in C[a,b], \tag{3.20}$$

is a finite sum

$$\tilde{\varphi}(x) := \sum_{i=1}^N x(t_i)w_i, \quad x \in C[a,b], \tag{3.21}$$

where $t_1, \ldots, t_N$ are distinct points in $[a,b]$ with

$$a \le t_1 < t_2 < \ldots < t_N \le b,$$

and $w_1, \ldots, w_N$ are real numbers. The points $t_1, \ldots, t_N$ in $[a,b]$ are called the **nodes** and $w_1, \ldots, w_N$ are called the **weights** of the quadrature rule $\tilde{\varphi}$.

Remark 3.5. It is easily seen that φ and $\tilde{\varphi}$ defined in (3.20) and (3.21), respectively, are continuous linear functionals on the Banach space $C[a,b]$ with respect to the norm $\| \cdot \|_\infty$, and $\|\varphi\| = b - a$ and $\|\tilde{\varphi}\| \le \sum_{i=1}^N |w_i|$. In fact, $\|\tilde{\varphi}\| = \sum_{i=1}^N |w_i|$. $\qquad \diamond$

Exercise 3.9. Justify the statements in Remark 3.5.

Suppose that for each positive integer n we have a quadrature rule φ_n determined by the nodes $t_1^{(n)}, \ldots, t_{k_n}^{(n)}$ in $[a,b]$ and weights $w_1^{(n)}, \ldots, w_{k_n}^{(n)}$, that is,

$$\varphi_n(x) := \sum_{i=1}^{k_n} x(t_i^{(n)})w_i^{(n)}, \quad x \in C[a,b]. \tag{3.22}$$

Here, (k_n) is an increasing sequence of positive integers. In this case, we may also say that (φ_n) is a quadrature rule. We say that (φ_n) is a **convergent quadrature rule** if

$$\varphi_n(x) \to \varphi(x) \quad \text{as} \quad n \to \infty$$

for every $x \in C[a,b]$.

Regarding the convergence of (φ_n) we have the following result.

Proposition 3.2. *The sequence (φ_n) in (3.22) of quadrature rules converges if and only if it converges on a dense subset of $C[a,b]$ and*

$$\sup_n \sum_{i=1}^{k_n} |w_i^{(n)}| < \infty.$$

Proof.   We know (cf. Remark 3.5) that

$$\|\varphi_n\| = \sum_{i=1}^{k_n} |w_i^{(n)}|$$

so that $\sup_n \sum_{i=1}^{k_n} |w_i^{(n)}| < \infty$ if and only if $(\|\varphi_n\|)$ is bounded. Now suppose that the sequence (φ_n) converges on $C[a,b]$. Then, by Uniform Boundedness Principle (Theorem 2.9), it follows that the sequence $(\|\varphi_n\|)$ is bounded.

Conversely, suppose that (φ_n) converges on a dense subset of $C[a,b]$ and $(\|\varphi_n\|)$ is bounded. Then by Theorem 2.7, (φ_n) converges on $C[a,b]$. $\square$

3.5.2   *Interpolatory quadrature rule*

Suppose P is an interpolatory projection given by

$$Px = \sum_{i=1}^{N} x(t_j)u_j, \quad x \in X,$$

where X is either $C[a,b]$ or $B[a,b]$ with norm $\|\cdot\|_\infty$ and $u_i \in X$, $i = 1,\ldots,N$, are Riemann integrable. Then, associated with P we have a quadrature rule, namely,

$$\tilde{\varphi}(x) = \int_a^b (Px)(t)dt = \sum_{i=1}^{N} x(t_i)w_i, \quad x \in C[a,b],$$

with

$$w_i = \int_a^b u_i(t)dt, \quad i \in \{1,\ldots,N\}.$$

Conversely, if we have a quadrature rule

$$\tilde{\varphi}(x) = \sum_{i=1}^{N} u(t_i)w_i, \quad x \in C[a,b],$$

then it is always possible to choose Riemann integrable functions (even continuous functions) $u_1,\ldots,u_N$ such that

$$u_i(t_j) = \begin{cases} 1 \text{ if } i = j \\ 0 \text{ if } i \neq j \end{cases} \quad \text{and} \quad \int_a^b u_j(t)dt = w_j,$$

so that P defined by $Px = \sum_{i=1}^{N} x(t_j)u_j$, $x \in X$, satisfies

$$\tilde{\varphi}(x) = \int_a^b (Px)(t)dt, \quad x \in C[a,b].$$

The above quadrature rule is called an **interpolatory quadrature rule** *induced by or associated with the interpolatory projection* P. It is to be observed that a quadrature rule can associate more than one interpolatory projections.

We note that if (P_n) is a sequence of interpolatory projections on $C[a, b]$ which converges pointwise to the identity operator I on $C[a, b]$, then

$$\varphi_n(x) = \int_a^b (P_n x)(t)dt \to \varphi(x), \quad x \in C[a, b].$$

Thus, we can say that an interpolatory quadrature rule associated with a sequence of pointwise convergent interpolatory projections converges.

We have observed in Section 3.4.1 that if $P_n x$ is the *Lagrange interpolation* of $x \in C[a, b]$ based on nodes $t_i^{(n)}$, $i = 1, \ldots, n$ with

$$a = t_0^{(n)} < t_1^{(n)} < \ldots < t_n^{(n)}$$

and

$$\max\{t_j^{(n)} - t_{j-1}^{(n)} : j = 1, \ldots, n\} \to 0 \quad \text{as} \quad n \to \infty,$$

then it is not necessary that $\|P_n u - u\|_\infty \to 0$. However, the corresponding quadrature rule can converge. In fact, this is the case if we take $t_1^{(n)}, \ldots, t_n^{(n)}$ as the zeros of the n-th *Legendre polynomial* (cf. [76] or [51]).

3.5.3 *Nyström approximation*

Consider the integral operator K in (3.14) with continuous kernel which is a compact operator on $C[a, b]$ (cf. Example 2.16). Let (φ_n) be a convergent quadrature rule on $C[a, b]$ given by

$$\varphi_n(x) := \sum_{i=1}^n x(t_i^{(n)})w_i^{(n)}, \quad x \in C[a, b].$$

Then the **Nyström approximation** associated with the quadrature rule (φ_n) is a sequence (K_n) of operators, where

$$(K_n x)(s) := \sum_{i=1}^n k(s, t_i^{(n)})x(t_i^{(n)})w_i^{(n)} \tag{3.23}$$

for $x \in C[a, b]$ and $s \in [a, b]$. We note that, for each $n \in \mathbb{N}$,

$$\|K_n x\|_\infty \leq \sup_{s,t\in\Omega} |k(s,t)| \sum_{i=1}^n |w_i^{(n)}| \, \|x\|_\infty \quad \forall \, x \in C[a, b],$$

so that $K_n \in \mathcal{B}(X)$. In fact, K_n is a finite rank operator with

$$R(K_n) \subseteq \text{span}\,\{k(\cdot, t_i^{(n)}) : i = 1, \ldots, n\}.$$

In particular, each K_n is a compact operator on $C[a, b]$.

Theorem 3.5 below shows that the Nyström approximation (K_n) is a pointwise approximation of K, however, we do not have the norm convergence (Theorem 3.6). For the proof of Theorem 3.5, we shall make use of the following lemma from real analysis which follows from Arzela-Ascoli theorem (Theorem 2.6). Here, we present an independent proof for the same.

Lemma 3.2. *Let (x_n) be an equicontinuous sequence in $C[a, b]$ and $x \in C[a, b]$ be such that for each $t \in [a, b]$, $|x_n(t) - x(t)| \to 0$ as $n \to \infty$. Then $\|x_n - x\|_\infty \to 0$ as $n \to \infty$.*

Proof.   Let $\varepsilon > 0$ be given. By the equicontinuity of (x_n), there exists $\delta_1 > 0$ such that

$$|x_n(t) - x_n(s)| < \varepsilon \quad \forall n \in \mathbb{N} \quad \text{whenever} \quad |s - t| < \delta_1. \qquad (3.24)$$

Since $x \in C[a, b]$ is uniformly continuous on $[a, b]$, there exists $\delta_2 > 0$ such that

$$|x(t) - x(s)| < \varepsilon \quad \text{whenever} \quad |s - t| < \delta_2. \qquad (3.25)$$

Let $\delta := \min\{\delta_1, \delta_2\}$. By the compactness of the interval $[a, b]$, there exist $t_1, \ldots, t_k$ such that $[a, b] \subseteq \cup_{i=1}^k \{s \in [a, b] : |s - t_i| < \delta\}$. For each $i \in \{1, \ldots, k\}$, let $N_i \in \mathbb{N}$ be such that

$$|x_n(t_i) - x(t_i)| < \varepsilon \quad \forall n \geq N_i. \qquad (3.26)$$

Now, let $t \in [a, b]$, and let $j \in \{1, \ldots, k\}$ be such that $|t - t_j| < \delta$. Thus, by (3.24), (3.25) and (3.26), we have

$$|x_n(t) - x(t)| \leq |x_n(t) - x_n(t_j)| + |x_n(t_j) - x(t_j)| + |x(t_j) - x(t)|$$
$$< 3\varepsilon$$

for all $n \geq N := \max\{N_j : j = 1, \ldots, k\}$. This is true for all $t \in [a, b]$. Hence,

$$\|x_n - x\|_\infty := \sup_{a \leq t \leq b} |x_n(t) - x(t)|$$
$$< 3\varepsilon$$

for all $n \geq N$. Thus, we have proved that $\|x_n - x\|_\infty \to 0$ as $n \to \infty$.   □

Remark 3.6. We may observe that in proving Lemma 3.2, the only fact that we used about $[a, b]$ is its compactness. Thus, Lemma 3.2 holds by replacing $[a, b]$ by any compact metric space Ω. $\Diamond$

Now, we prove the pointwise convergence of (K_n) to K.

Theorem 3.5. *Let* $X = C[a, b]$ *with* $\| \cdot \|_\infty$ *and* (K_n) *be the Nyström approximation of the integral* K *in (3.14) with continuous kernel* $k(\cdot, \cdot)$. *Then the following hold:*

(i) $\cup_{n=1}^{\infty}\{K_n x : x \in C[a, b] \, \|x\|_\infty \leq 1\}$ *is equicontinuous.*

(ii) (K_n) *is a pointwise approximation of* K.

Proof.
Writing $k_s(t) = k(s, t)$, by the convergence of the quadrature rule φ_n, we have

$$(K_n x)(s) = \varphi_n(k_s x) \to \varphi(k_s x) = (Kx)(s)$$

for each $x \in C[a, b]$ and $s \in [a, b]$. Moreover for $s, \tau \in [a, b]$ and $x \in C[a, b]$ with $\|x\|_\infty \leq 1$,

$$|(K_n x)(s) - (K_n x)(\tau)| \leq c \sup_{a \leq t \leq b} |k(s, t) - k(\tau, t)|,$$

where $c \geq \sum_{i=1}^{n} |w_i^{(n)}|$. Hence, by uniform continuity of $k(\cdot, \cdot)$, the set

$$\{K_n x : x \in C[a, b] \text{ with } \|x\|_\infty \leq 1, \, n \in \mathbb{N}\}$$

and, for each $x \in C[a, b]$, the set $\{K_n x : n \in \mathbb{N}\}$ are equicontinuous. Since we also have

$$|(Kx)(s) - (K_n x)(s)| \to 0 \quad \text{as} \quad n \to \infty$$

for each $s \in [a, b]$, by Lemma 3.2, it follows that $(K_n x)$ converges uniformly to Kx for each $x \in C[a, b]$. Thus,

$$\|Kx - K_n x\|_\infty \to 0 \quad \text{as} \quad n \to \infty$$

for each $x \in C[a, b]$. Thus we have proved (i) and (ii). $\square$

Corollary 3.4. *Let* K *and* K_n *be as in (3.14) and (3.23) respectively. Then*

$$\|(K - K_n)K\| \to 0 \quad and \quad \|(K - K_n)K_n\| \to 0.$$

as $n \to \infty$.

Proof. By Theorem 3.5, (K_n) is a pointwise approximation of the compact operator K and the set

$$S := \{K_n x : x \in C[a,b] \text{ with } \|x\|_\infty \le 1, n \in \mathbb{N}\}$$

is equicontinuous. Hence, by Theorem 2.13, $\|(K - K_n)K\| \to 0$ and

$$\|(K - K_n)K_n\| \le \sup_{u \in S} \|(K - K_n)u\| \to 0$$

as $n \to \infty$.   □

Next theorem shows that the Nyström approximation is not a norm approximation.

Theorem 3.6. *Let $X = C[a,b]$ with $\| \cdot \|_\infty$ and (K_n) be the Nyström approximation of the integral K in (3.14) with continuous kernel $k(\cdot, \cdot)$. Then*

$$\|K - K_n\| \ge \|K\| \quad \forall n \in \mathbb{N}.$$

Proof. Let $\varepsilon > 0$ and Ω_ε and $x_\varepsilon \in C[a,b]$ be as in Lemma 3.1, satisfying the conditions as in Remark 3.4, namely,

$$t_i^{(n)} \in \Omega_\varepsilon \quad \text{and} \quad x_\varepsilon(t_i^{(n)}) = 0, \quad i = 1, \ldots, n.$$

Then we have

$$K_n x_\varepsilon = 0 \quad \forall n \in \mathbb{N}$$

and

$$\|K - K_n\| \ge \|(K - K_n)x_\varepsilon\|_\infty = \|K x_\varepsilon\|_\infty.$$

Hence, by Lemma 3.1, we have $\|K - K_n\| \ge \|K\|$ for all $n \in \mathbb{N}$.   □

3.5.4   *Collectively compact approximation*

The properties of Nyström approximation (K_n) enlisted in Theorem 3.5 have been observed long ago by Brackage [9]. These properties have been studied by Anselone [3] in the operator theoretic setting, and called such a sequence of operators *collectively compact*.

A sequence (K_n) in $\mathcal{B}(X,Y)$ is said to be **collectively compact** if closure of the set

$$\bigcup_{n=k}^{\infty} \{K_n x : \|x\| \le 1\}$$

is compact in Y for some $k \in \mathbb{N}$.

A very good reference, apart from the book of Anselone [3], where collectively compact approximation has been dealt for the solution of integral equation of the second kind is the recent book by Atkinson [5].

The following theorem can be proved easily.

Theorem 3.7. *Suppose* (K_n) *is a collectively compact sequence in* $\mathcal{B}(X,Y)$. *Then the following hold.*

(i) *Each* K_n *is compact.*

(ii) *If* $K_n \to K$ *pointwise, then* K *is compact.*

Exercise 3.10. Prove Theorem 3.7.

In order to apply the property of collectively compactness to possibly non-compact operators, we consider the following definition, as in Chatelin [10] and Limaye [38].

A sequence (T_n) in $\mathcal{B}(X,Y)$ is said to be a **collectively compact approximation** of $T \in \mathcal{B}(X,Y)$ if $T_n \to T$ pointwise and $(T_n - T)$ is a collectively compact sequence.

By Theorem 3.7, if (E_n) is a sequence of non-compact operators such that $\|E_n\| \to 0$ as $n \to \infty$, then (T_n) with $T_n = T + E_n$ converges to T in norm, but not in a collectively compact manner. For example, we may take $Y = X$ and $E_n = (1/n)I$. Thus, the class of collectively compact approximations does not include the class of norm approximations. However, if $T_n - T$ are compact operators for all large enough n, then norm convergence does imply collectively compact convergence (cf. [38]).

In particular, we have the following result.

Theorem 3.8. *Suppose that* (K_n) *is a sequence in* $\mathcal{K}(X)$ *which is a norm approximation of some* $K \in \mathcal{B}(X)$. *Then* (K_n) *is a collectively compact approximation of* K.

Remark 3.7. We know from Theorem 2.17 that a compact operator on an infinite dimensional Banach space can never be bijective. However, Theorem 3.8 is still useful for a well-posed equation (3.1). For instance, consider the second kind equations (3.27) and (3.28), that is,

$$\lambda x - Kx = y, \quad \lambda x_n - K_n x_n = y_n$$

with K and K_n being compact operators on X. In this case, we have $T = I - K$ and $T_n = I - K_n$ so that $T_n = T + (K - K_n)$, a compact perturbation of T. It follows from Theorem 3.8 that if $\|K - K_n\| \to 0$ then (K_n) is a collectively compact approximation of K, and hence (T_n) is a collectively compact approximation of T. $\diamond$

3.6   Norm-square Approximation Revisited

In order to be able to apply Theorem 3.4 in Section 3.3.2, to the situations described in the last two subsections, we consider the well-posed operator equation (3.1) with $T = \lambda I - K$, i.e.,

$$\lambda x - Kx = y \tag{3.27}$$

where $K \in \mathcal{B}(X)$ and λ is a nonzero scalar not in the spectrum of K. Suppose (K_n) is a sequence in $\mathcal{B}(X)$ and (y_n) is a sequence in Y. Consider the associated approximate equation (3.2) with $T_n = \lambda I - K_n$, i.e.,

$$\lambda x_n - K_n x = y_n. \tag{3.28}$$

Throughout this section we assume that X is a Banach space.

Clearly, $(\lambda I - K_n)$ is a norm-square approximation of $\lambda I - K$ if and only if

(A0) $\|[(K - K_n)(\lambda I - K)^{-1}]^2\| \to 0.$

Let us consider two conditions which are independent of λ:

(A1) $\|(K - K_n)K\| \to 0.$

(A2) $\|(K - K_n)K_n\| \to 0.$

The above conditions (A1) and (A2) which are used in Ahues [1], Bouldin [7] and Nair [44] in the context of spectral approximations have been made use extensively in the book [2] by Ahues, Largillier and Limaye. In [2], the convergence of (K_n) to K in the sense that the conditions (A1) and (A2) together with the boundedness of $(\|K_n\|)$ are satisfied is called the *ν-convergence*.

Remark 3.8. By Corollary 3.4, if (K_n) is a Nyström approximation of the integral operator in (3.14), then the conditions (A1) and (A2) are satisfied. In this case, by Theorem 3.6, (K_n) is not a norm approximation of K. $\diamond$

Next, our aim is to show that conditions (A1) and (A2) together with boundedness of (K_n) imply norm-square convergence. In this regard, we consider another condition on (K_n):

(A3) $\|(K - K_n)^2\| \to 0.$

Note that if (K_n) is a norm approximation of K, then all the conditions (A0)-(A3) are satisfied.

The following theorem is a consequence of the identity

$$(K - K_n)K = (K - K_n)K_n + (K - K_n)^2.$$

Theorem 3.9. *Any two of (A1)-(A3) imply the third.*

Next two theorems specify certain class of approximations which satisfy conditions (A1)-(A3).

Theorem 3.10. *Suppose (P_n) is a sequence of projections in $\mathcal{B}(X)$ such that $(\|P_n\|)$ is bounded and $\|(I - P_n)K\| \to 0$, and let*

$$K_n \in \{P_nK, KP_n, P_nKP_n\}.$$

Then (K_n) satisfies (A1)-(A3).

Proof.   Obvious from the assumptions.   □

Theorem 3.11. *Suppose (K_n) is a collectively compact approximation of K. Then (A3) is satisfied. If in addition K is compact, then condition (A1)-(A2) are also satisfied.*

Proof.   By the assumption on (K_n), the closure of the set

$$S := \bigcup_{n=1}^{\infty} \{Kx - K_nx : x \in X, \|x\| \le 1\}$$

is compact in Y. Hence, by Theorem 2.13,

$$\|(K - K_n)^2\| \le \sup_{x \in S} \|(K - K_n)x\| \to 0 \quad \text{as} \quad n \to \infty$$

so that (A3) is satisfied. Next, suppose that K is compact. Then, again by Theorem 2.13,

$$\|(K - K_n)K\| \to 0 \quad \text{as} \quad n \to \infty$$

so that (A1) is satisfied, and consequently, by Theorem 3.9, (A2) is also satisfied.   □

Next we link the conditions (A0) with (A1)-(A3).

Theorem 3.12. *Suppose* $(\|K_n\|)$ *is bounded. Then any two of the conditions (A1)-(A3) imply (A0).*

Proof.   By Theorem 3.9, any two of (A1)-(A3) imply the third. Hence, it is enough to show that conditions (A1) and (A3) imply (A0). For this, first we observe that for every nonzero scalar $\lambda \notin \sigma(K)$,

$$(\lambda I - K)^{-1} = \frac{1}{\lambda}[I + K(\lambda I - K)^{-1}].$$

Thus, denoting $R_\lambda := (\lambda I - K)^{-1}$ and $T_\lambda := I + K(\lambda I - K)^{-1}$, we have

$$\lambda^2[(K - K_n)R_\lambda]^2 = [(K - K_n)T_\lambda]^2$$
$$= [(K - K_n)^2 + (K - K_n)KR_\lambda(K - K_n)]T_\lambda.$$

From this, it is clear that conditions (A1) and (A3) together with the boundedness of $(\|K_n\|)$ imply (A0). $\qquad\square$

Corollary 3.5. *Suppose K is a compact operator and (K_n) is a pointwise approximation of K. Then $(\lambda I - K_n)$ is a norm-square approximation of $\lambda I - K$ if and only if $\|(K - K_n)^2\| \to 0$ as $n \to \infty$.*

Proof.   Since (K_n) is a pointwise approximation of K, by Banach Steinhaus theorem (Theorem 2.11), $(\|K_n\|)$ is bounded. Also, since K is compact, Theorem 2.13 implies that $\|(K_n - K)K\| \to 0$ as $n \to \infty$. Now, the result is a consequence of Theorem 3.12. $\qquad\square$

The following example shows that the condition (A0) is (strictly) weaker than the conditions required for ν-convergence, namely the conditions (A1) and (A2).

Example 3.7. Let $X = \ell^2$ and A_n be as in Example 3.6. Let

$$K = I, \quad \lambda \neq 1 \quad \text{and} \quad K_n := I - A_n.$$

Then we have

$$\|(K - K_n)K\| = \|A_n\| = 1$$

$$\|(K - K_n)K_n\| = \|A_n(I - A_n)\| \geq \|A_n\| - \|A_n^2\| = 1 - \frac{1}{n^2}$$

for all $n \in \mathbb{N}$, whereas

$$\|[(K - K_n)(K - \lambda I)^{-1}]^2\| = \frac{\|A_n^2\|}{(1 - \lambda)^2} = \frac{1}{n^2(1 - \lambda)^2} \to 0$$

as $n \to \infty$. Thus, conditions (A1) and (A2) are not satisfied, but (A0) is satisfied. $\qquad\diamond$

The following theorem is a consequence of Theorems 3.4 and 3.12. However, we give an alternate proof for the same.

Theorem 3.13. *Let (K_n) be a uniformly bounded sequence in $\mathcal{B}(X)$ such that any two of the conditions (A1)–(A3) are satisfied. Then $(\lambda I - K_n)$ is stable.*

Proof. [An alternate proof.] In view of Theorem 3.9 we assume conditions (A2) and (A3), that is,
$$\|(K - K_n)K_n\| \to 0, \quad \|(K - K_n)^2\| \to 0.$$
For $n \in \mathbb{N}$, let
$$A_n := \lambda I - (K - K_n), \quad B_n := \lambda I + (\lambda I - K)^{-1}(K - K_n)K_n.$$
First we observe the identity
$$A_n(\lambda I - K_n) = (\lambda I - K)B_n. \tag{3.29}$$
Let $0 < c_0 < 1$ and N be a positive integer such that
$$\|(K - K_n)^2\| \le (c_0|\lambda|)^2$$
and
$$\|(K - K_n)K_n\| \le \frac{c_0|\lambda|}{\|(\lambda I - K)^{-1}\|}$$
for all $n \ge N$. Then, by Theorem 2.22 and Corollary 2.14, the operators A_n and B_n are invertible for all $n \ge N$. Therefore, from (3.29), it follows that $\lambda I - K_n$ is invertible for all $n \ge N$, and
$$(\lambda I - K_n)^{-1} = B_n^{-1}(\lambda I - K)^{-1}A_n.$$
This leads to the inequality
$$\|(\lambda I - K_n)^{-1}\| \le \frac{\|(\lambda I - K)^{-1}\|(|\lambda| + \|K - K_n\|)}{|\lambda| - \|(\lambda I - K)^{-1}\|\|(K - K_n)K_n\|}.$$
Now let $M > 0$ be such that $\|K - K_n\| \le |\lambda|M$ for every $n \in \mathbb{N}$. Such an M exists since $(\|K_n\|)$ is bounded. Hence from the above inequality we have
$$\|(\lambda I - K_n)^{-1}\| \le \frac{1 + M}{1 - c_0}\|(\lambda I - K)^{-1}\| \quad \forall n \ge N.$$
This completes the proof. $\qquad\square$

Now, in view of Theorem 3.12 and Corollary 3.5, we have the following.

Corollary 3.6. *The conclusions of Theorem 3.13 hold if one of the following is satisfied.*

(i) *(K_n) is a norm approximation of K.*

(ii) *K is a compact operator, (K_n) is a pointwise approximation of K and $\|(K - K_n)^2\| \to 0$.*

Exercise 3.11. Write detailed proof of the above corollary.

3.7   Second Kind Equations

As we have already remarked earlier, a typical form of well-posed equations which often occur in applications is the equation of the second kind,

$$\lambda x - Kx = y, \tag{3.30}$$

where $K : X \to X$ is a bounded operator on a Banach space X, $y \in X$ and λ is a nonzero scalar which is not a spectral value of K. This equation is a particular case of (3.1) in which $T = \lambda I - K$. Recall from Theorem 2.25 that if K is a compact operator, then the requirement on λ that it is not a spectral value is the same as saying that it is not an eigenvalue of K. In fact, this special case is important, as mathematical formulation of many well-posed equations in applications appear in the form of integral equations of the second kind,

$$\lambda x(s) - \int_{\Omega} k(s,t)x(t)dt = y(s), \quad s \in \Omega, \tag{3.31}$$

where the kernel $k(\cdot, \cdot)$ and the set $\Omega \subseteq \mathbf{R}^k$ are such that the operator K defined by

$$(Kx)(s) = \int_{\Omega} k(s,t)x(t)dt, \quad s \in \Omega, \tag{3.32}$$

is a compact operator on a suitable function space. In Chapter 2, we have discussed many such examples. To see the compactness of the above integral operator in more general contexts the reader may refer the book by Kress [35] and the paper by Graham and Sloan [22].

As we have already discussed in the last section, in order to approximate the solution of (3.30), one may have a sequence (K_n) in $\mathcal{B}(X)$ and (y_n) in Y instead of K and y. Thus the equation at hand would be

$$\lambda x_n - K_n x_n = y_n. \tag{3.33}$$

For instance, we may have a norm approximation (K_n) of K, that is

$$\|K - K_n\| \to 0 \quad \text{as} \quad n \to \infty$$

so that by taking $T = \lambda I - K$ and $T_n = \lambda I - K_n$ we have

$$\|T - T_n\| \to 0 \quad \text{as} \quad n \to \infty.$$

One such case is when $K_n := P_n K$, where K is a compact operator on X and (P_n) is a sequence of projections in $\mathcal{B}(X)$ which is a pointwise approximation of the identity operator. Recall that if K is a Fredholm integral operator on $C[a, b]$ and (K_n) is a Nyström approximation of K, then (K_n) is *not a norm approximation* of K (cf. Theorem 3.6). However, (K_n) is a pointwise approximation of K and

$$\|(K - K_n)K\| \to 0, \quad \|(K - K_n)K_n\| \to 0$$

as $n \to \infty$ (Theorem 3.6 and Corollary 3.4). Hence, by Theorem 3.13, Theorem 3.4 can be applied with $T := \lambda I - K$ and $T_n := \lambda I - K_n$.

We note that if x and x_n are solutions of (3.30) and (3.33), respectively, then (3.8) takes the form

$$(\lambda I - K_n)(x - x_n) = (K - K_n)x + (y - y_n). \tag{3.34}$$

The following theorem is a consequence of Theorems 3.3 and 3.13.

Theorem 3.14. *Assume that either* $\|K - K_n\| \to 0$ *or any two of the conditions (A1)–(A3) are satisfied. Then* $(\lambda I - K_n)$ *is stable with stability index N and*

$$c_1 \alpha_n \leq \|x - x_n\| \leq c_2 \alpha_n \quad \forall n \geq N,$$

where x and x_n are solutions of (3.30) and (3.33), respectively, $\alpha_n :=$ $\|(K - K_n)x + (y - y_n)\|$, *$c_1$ and c_2 are positive real numbers independent of n.*

Exercise 3.12. Write detailed proof.

3.7.1 *Iterated versions of approximations*

After obtaining an approximation x_n as the solution of the equation (3.33) for the solution x of (3.30), one may look for a better approximation, that is, an $\tilde{x}_n$ having better rate of convergence than x_n. In this connection, the first attempt would be to look for an iterated version of x_n, obtained by replacing x on the right-hand side of the identity

$$x = \frac{1}{\lambda}(y + Kx)$$

by x_n, that is, by defining $\tilde{x}_n$ as

$$\tilde{x}_n = \frac{1}{\lambda}(y + Kx_n). \tag{3.35}$$

Then we have

$$x - \tilde{x}_n = \frac{1}{\lambda}K(x - x_n) \tag{3.36}$$

and obtain the following theorem on error estimate.

Theorem 3.15. *Let x, x_n be the solutions of (3.30) and (3.33), respectively, and $\tilde{x}_n$ be defined by (3.35). Then*

$$\|x - \tilde{x}_n\| \leq \frac{1}{|\lambda|} \|K(x - x_n)\| \qquad (3.37)$$

and

$$c_{1,\lambda} \|\tilde{x}_n - x_n\| \leq \|x - x_n\| \leq c_{2,\lambda} \|\tilde{x}_n - x_n\|, \qquad (3.38)$$

hold, where

$$c_{1,\lambda} = \frac{1}{\|\lambda(\lambda I - K)\|} \quad and \quad c_{2,\lambda} = \frac{\|(\lambda I - K)^{-1}\|}{|\lambda|}.$$

Proof.   The estimate in (3.37) follows from (3.36). To obtain the estimate in (3.38), first we observe that

$$\tilde{x}_n - x_n = \frac{1}{\lambda}\{(y - y_n) + (K - K_n)x_n\}.$$

Hence, estimate in (3.38) is obtained from (3.10) since the quantity β_n there takes the form

$$\beta_n = \|(y - y_n) + (K - K_n)x_n\| = |\lambda| \|\tilde{x}_n - x_n\|.$$

This completes the proof.   □

By the estimate in (3.37), the accuracy of $\tilde{x}_n$ is at least that of x_n. We can say that the rate of convergence of $(\tilde{x}_n)$ is higher than that of (x_n) provided there exists a sequence (η_n) of positive reals such that $\eta_n \to 0$ as $n \to \infty$ and

$$\|K(x - x_n)\| \leq \eta_n \|x - x_n\|.$$

We shall describe such cases in due course. We may also observe that the bounds in (3.38) is purely computational, not involving the 'unknown' x. We shall discuss this case again in the context of projection methods and quadrature methods for integral equations.

3.8   Methods for Second Kind Equations

We have already given conditions on the approximating operators K_n so that equation (3.33) is uniquely solvable for all large enough n, and also obtained certain estimates for the error $\|x - x_n\|$, where x is the unique solution of equation (3.30).

In practical situations, the approximating operators K_n are of finite rank, say $R(K_n) \subseteq X_n$ with $\dim(X_n) < \infty$, so that, as explained in Section 3.1, x_n can be obtained using a finite system of linear algebraic equations. Having obtained the solution x_n of (3.33), one would like to modify the method so as to have a better order of convergence and/or better convergence properties. The modified approximation may be an equation of the second kind, but not necessarily in a finite dimensional setting.

In this section we shall consider the above aspects for two types of methods. The first one is based on a sequence of projections on X and the second is based on quadrature rules for the particular case of the integral equation (3.31).

3.8.1 *Projection methods*

We would like to obtain approximations for the solution of equation (3.30) with the help of a sequence (P_n) of projections in $\mathcal{B}(X)$. In applications, the projections P_n are usually of finite rank, though our analysis of the methods often does not require this restriction.

Throughout this subsection we assume that $K \in \mathcal{B}(X)$ with X being a Banach space, λ is a nonzero scalar not belonging to the spectrum of K and the sequence (P_n) of projections in $\mathcal{B}(X)$ satisfies

$$\|(I - P_n)K\| \to 0 \quad \text{as} \quad n \to \infty. \tag{3.39}$$

Hence, by Theorem 3.14, $(\lambda I - P_n K)$ is stable.

Recall that if (P_n) converges pointwise to the identity operator and K is a compact operator, then we do have the convergence in (3.39). It is easily seen that if (X_n) is a sequence of finite dimensional subspaces of X such that

$$X_n \subseteq X_{n+1} \quad \text{and} \quad \text{cl} \bigcup_{n=1}^{\infty} X_n = X,$$

then for each $n \in \mathbb{N}$ we can define a projection P_n onto X_n such that for all $x \in X$,

$$\|x - P_n x\| \to 0 \quad \text{as} \quad n \to \infty.$$

Exercise 3.13. Using the assumptions, prove the conclusion in the last statement.

In the following we shall consider essentially four methods, namely the (i) Galerkin method, (ii) iterated Galerkin method, (iii) Kantorovich method and (iv) a modified projection method. We shall also discuss some iterated versions of the above methods.

(i) Galerkin method

The idea in Galerkin method is to find a sequence (x_n) in X, for large enough n, such that

$$x_n \in X_n := R(P_n) \quad \text{and} \quad P_n(\lambda x_n - K x_n - y) = 0.$$

Equivalently, one has to solve the equation

$$\lambda x_n - P_n K x_n = P_n y. \tag{3.40}$$

Note that this is of the form (3.33) with $y_n = P_n y$ and

$$K_n = P_n K \quad \text{or} \quad K_n = P_n K P_n.$$

Suppose P_n is of finite rank. In the following discussion, for the sake of simplicity, we assume $\text{rank}\,(P_n) = n$. Then we know that P_n can be represented by

$$P_n u = \sum_{j=1}^{n} f_j^{(n)}(u) u_j^{(n)}, \quad u \in X, \tag{3.41}$$

where $\{u_1^{(n)}, \ldots, u_n^{(n)}\}$ is a basis of $X_n := R(P_n)$ and $f_1^{(n)}, \ldots, f_n^{(n)}$ are continuous linear functionals on X such that

$$f_i^{(n)}(u_j^{(n)}) = \begin{cases} 1 \text{ if } i = j \\ 0 \text{ if } i \neq j. \end{cases} \tag{3.42}$$

For the sake of simplicity of presentation, at the cost of preciseness, we shall omit the superscripts in $u_1^{(n)}, \ldots, u_n^{(n)}$ and $f_1^{(n)}, \ldots, f_n^{(n)}$, and write as $u_1, \ldots, u_n$ and $f_1, \ldots, f_n$, respectively.

The following theorem shows that the problem of solving (3.40) is equivalent to that of solving a system of equations involving scalar variables.

Theorem 3.16. *Let P_n be as in (3.41) and $(\alpha_1, \ldots, \alpha_n) \in \mathbb{K}^n$. Then*

$$x_n = \sum_{j=1}^{n} \alpha_j u_j$$

is a solution of (3.40) if and only if

$$\lambda \alpha_i - \sum_{j=1}^{n} a_{ij} \alpha_j = f_i(y), \quad i = 1, \ldots, n, \tag{3.43}$$

where $a_{ij} = f_i(K u_j)$ for $i, j \in \{1, \ldots, n\}$.

Proof. First we observe, using the representation (3.41) of P_n, that equation (3.40) takes the form

$$\lambda x_n - \sum_{j=1}^{n} f_j(Kx_n)u_j = \sum_{i=1}^{n} f_j(y)u_j.$$

Now suppose that $x_n = \sum_{j=1}^{n} \alpha_j u_j$ is a solution of (3.40). Then applying f_i to the above equation and using the property (3.42), we get

$$\lambda f_i(x_n) - f_i(Kx_n) = f_i(y).$$

But

$$f_i(x_n) = \sum_{j=1}^{n} \alpha_j f_i(u_j) = \alpha_i$$

and

$$f_i(Kx_n) = f_i\Big(\sum_{j=1}^{n} \alpha_j Ku_j\Big) = \sum_{j=1}^{n} \alpha_j f_i(Ku_j).$$

Thus we see that $(\alpha_1, \ldots, \alpha_n) \in \mathbb{K}^n$ satisfies (3.43).

Conversely, suppose that $(\alpha_1, \ldots, \alpha_n) \in \mathbb{K}^n$ satisfies (3.43) and let $x_n = \sum_{j=1}^{n} \alpha_j u_j$. Now, (3.43) implies that

$$\lambda \sum_{i=1}^{n} \alpha_i u_i - \sum_{i=1}^{n} \Big(\sum_{j=1}^{n} f_i(Ku_j)\alpha_j \Big) u_i = \sum_{i=1}^{n} f_i(y)u_i$$

which is the same as (3.40) with $x_n = \sum_{j=1}^{n} \alpha_j u_j$. $\qquad \square$

We obtain different methods by choosing different type of projections, which is decided by the choice of the basis elements u_j's and the functionals f_j's. This choice is often dictated by the computational feasibility of the method.

The terminology *Galerkin method*, named after the Russian mathematician Galerkin (Boris Grigoryvich Galerkin) (1871–1945), pronounced as "Gulyorkin", is often used when the space X is endowed with an inner product, which does not necessarily induce the original norm on X, and the projections involved are orthogonal projections with respect to this inner product. We call such method as *orthogonal Galerkin method*. We shall consider this special case and also another important procedure known as *collocation method* in the context of $X = C[a, b]$. These methods are general enough to cover various problems which one encounters in applications.

The solution x_n of equation (3.40) or the sequence (x_n) is called the **Galerkin approximation** of the solution x of (3.30). We shall denote the Galerkin approximation by x_n^G. Thus, x_n^G satisfies the equation

$$\lambda x_n^G - P_n K x_n^G = P_n y. \tag{3.44}$$

Hence, equation (3.34) gives

$$x - x_n^G = \lambda(\lambda I - P_n K)^{-1}(x - P_n x). \tag{3.45}$$

Theorem 3.17. *Let N be a stability index of $(\lambda I - P_n K)$. Then the following hold for all $n \geq N$.*

(i) $c_1 \|x - P_n x\| \leq \|x - x_n^G\| \leq c_2 \|x - P_n x\|.$

(ii) $\|x - x_n^G\| \to 0 \iff \|y - P_n y\| \to 0.$

(iii) *If $(\|P_n\|)$ is bounded, then*

$$dist(x, X_n) \leq \|x - x_n^G\| \leq c_3 \, dist(x, X_n).$$

Proof. The result in (i) follows from (3.45). The result in (ii) is a consequence of (i), since

$$(I - P_n)x = \frac{1}{\lambda}(I - P_n)(y + Kx)$$

and $\|(I - P_n)K\| \to 0$. To obtain (iii), suppose $(\|P_n\|)$ is bounded, say $\|P_n\| \leq c$ for all $n \in \mathbb{N}$. We first observe that

$$x_n^G = \frac{1}{\lambda}(P_n y + P_n K x_n^G) \in R(P_n).$$

Hence, it is clear that $dist(x, X_n) \leq \|x - x_n^G\|$. Also, for all $u_n \in X_n$,

$$\|x - P_n x\| = \|(I - P_n)(x - u_n)\| \leq (1 + \|P_n\|)\|x - u_n\|$$

so that

$$\|x - P_n x\| \leq (1 + c) \, dist(x, X_n).$$

This together with (i) gives

$$\|x - x_n^G\| \leq c_2\|x - P_n x\|\| \leq c_2(1 + c) \, dist(x, X_n).$$

Thus, (iii) holds with $c_3 = c_2(1 + c)$. $\square$

Remark 3.9. The result (iii) in Theorem 3.17 shows that the order of convergence of the Galerkin approximation (x_n^G) cannot be improved by any other element from $X_n := R(P_n)$. $\Diamond$

In view of the above remark, we say that a method which gives an approximation (z_n) is a **superconvergent method** if the corresponding order of convergence is better than $O(\|x - x_n^G\|)$, that is, if there exists a sequence (η_n) of positive reals such that $\eta_n \to 0$ and

$$\|x - z_n\| \leq \eta_n \|x - x_n^G\|$$

for all large n.

Our next attempt is to modify the Galerkin method so as to obtain a superconvergent method, at least in certain special cases. We shall see if the iterated Galerkin approximation provides superconvergence. Before that let us consider two special cases of Galerkin method, namely, the *orthogonal Galerkin method* and the *collocation method* in the context of $X = C[a, b]$.

(a) Orthogonal Galerkin method

Suppose the space X is also endowed with an inner product, say $\langle \cdot, \cdot \rangle_0$ such that the associated norm $\| \cdot \|_0$ is weaker than the original norm $\| \cdot \|$ on X, that is, there exists a constant $c > 0$ such that

$$\|x\|_0 \leq c \|x\| \quad \forall x \in X. \tag{3.46}$$

For each $n \in \mathbb{N}$, let X_n be a finite dimensional subspace of X and let $\{u_1, \ldots, u_n\}$ be a basis of X_n which is orthonormal with respect to the inner product $\langle \cdot, \cdot \rangle_0$, that is

$$\langle u_i, u_j \rangle_0 = \begin{cases} 1 \text{ if } i = j \\ 0 \text{ if } i \neq j. \end{cases}$$

In *orthogonal Galerkin method* one looks for $x_n \in X_n$ such that

$$\langle \lambda x_n - K x_n, u \rangle_0 = \langle y, u \rangle_0 \quad \forall u \in X_n,$$

which is equivalent to

$$\langle \lambda x_n - K x_n, u_i \rangle_0 = \langle y, u_i \rangle_0 \quad i = 1, \ldots, n. \tag{3.47}$$

Writing $x_n = \sum_{j=1}^{n} \alpha_j u_j$, the above system of equations is equivalent to the problem of finding $(\alpha_1, \ldots, \alpha_n) \in \mathbb{K}^n$ such that

$$\lambda \alpha_i - \sum_{j=1}^{n} a_{ij} \alpha_j = \beta_i, \quad i = 1, \ldots, n$$

with $a_{ij} = \langle K u_j, u_i \rangle_0$ and $\beta_i = \langle y, u_i \rangle_0$ for $i, j = 1, \ldots, n$. Define

$$P_n x = \sum_{j=1}^{n} \langle x, u_j \rangle_0 u_j, \quad x \in X.$$

It is seen that P_n is an orthogonal projection, and (3.47) is the same as equation (3.40). Clearly, P_n is continuous with respect to the norm $\| \cdot \|_0$. In view of the inequality (3.46), it also follows that $P_n \in \mathcal{B}(X)$.

Exercise 3.14. Justify the last two statements.

As an example of the above situation, one may take $X = C[a, b]$ with $\| \cdot \|_\infty$ or $X = L^2[a, b]$. In both cases we may take the inner product as

$$\langle x, u \rangle_0 := \int_a^b x(t)\overline{u(t)}\, dt, \quad x, y \in X.$$

Recall that $C[a, b]$ is not a Banach space with respect to $\| \cdot \|_0$.

(b) Collocation method

This method is specially meant for a Banach space X of functions defined on a subset Ω of $\mathbb{R}^k$. For the sake of simplicity of presentation, we take $X = C[a, b]$ with $\| \cdot \|_\infty$. In this method, we take P_n to be an interpolatory projection based on nodes $t_1^{(n)}, \ldots, t_n^{(n)}$ in $[a, b]$ and functions $u_1^{(n)}, \ldots, u_n^{(n)}$ in X satisfying

$$u_i^{(n)}(t_j v) = \begin{cases} 1 \text{ if } i = j \\ 0 \text{ if } i \neq j, \end{cases}$$

that is,

$$P_n x = \sum_{j=1}^n x(t_j^{(n)}) u_j^{(n)}, \quad x \in C[a, b]. \tag{3.48}$$

Since

$$(P_n x)(t_i^{(n)}) = x(t_i^{(n)}), \quad i = 1, \ldots, n,$$

the problem of solving equation (3.40) for $x_n = \sum_{j=1}^n \alpha_j u_j^{(n)}$ is same as the problem of finding $(\alpha_1, \ldots, \alpha_n) \in \mathbb{K}^n$ such that

$$\alpha_i - \sum_{j=1}^n a_{ij}\alpha_j = \beta_i, \quad i = 1, \ldots, n$$

with $a_{ij} = (Ku_j)(t_i^{(n)})$ and $\beta_i = y(t_i^{(n)})$ for $i, j = 1, \ldots, n$.

(ii) Iterated Galerkin method

Having defined Galerkin approximation x_n^G, its iterated version, introduced and studied extensively by Sloan (see e.g. [72] or [73]) and his collaborators, is defined by

$$x_n^S = \frac{1}{\lambda}(y + Kx_n^G).$$

Then x_n^S or the sequence (x_n^S) is called the **iterated Galerkin approximation** or **Sloan approximation** of x. As in (3.36) we have

$$x - x_n^S = \frac{1}{\lambda}K(x - x_n^G). \tag{3.49}$$

From the definition of the Galerkin approximation, it follows that

$$P_n x_n^S = \frac{1}{\lambda}(P_n y + P_n K x_n^G) = x_n^G. \tag{3.50}$$

If P_n is as in collocation method, (3.48), then the above observation implies that

$$x_n^S(t_i^{(n)}) = x_n^G(t_i^{(n)}), \quad i = 1, \ldots, n. \tag{3.51}$$

The observation (3.50) shows that

$$\begin{aligned} x_n^S &= \frac{1}{\lambda}(y + Kx_n^G) \\ &= \frac{1}{\lambda}(y + KP_n x_n^S). \end{aligned}$$

Thus, x_n^S satisfies a second kind equation of the form (3.33), namely

$$\lambda x_n^S - KP_n x_n^S = y. \tag{3.52}$$

Hence, equation (3.34) gives

$$(\lambda I - KP_n)(x - x_n^S) = K(x - P_n x). \tag{3.53}$$

From the above equation we also obtain

$$x - x_n^S = (\lambda I - KP_n)^{-1}K(I - P_n)(x - x_n^G). \tag{3.54}$$

The following lemma is a consequence of Theorems 3.10 and 3.13. However, we shall supply an independent proof for the same.

Lemma 3.3. *In addition to (3.39), assume that $(\|P_n\|)$ is bounded. Then $(\lambda I - KP_n)$ is stable.*

Proof. [An alternate proof.] By the assumption of (3.39), we know that the sequence $(\lambda I - P_n K)$ is stable. Now, by Theorem 2.23, the operators $P_n K$ and $K P_n$ have the same nonzero spectral values, and

$$\lambda(\lambda I - K P_n)^{-1} = I + K(\lambda I - P_n K)^{-1} P_n \qquad (3.55)$$

for all $n \geq N$. Hence, the result follows by the additional assumption that $(\|P_n\|)$ is bounded. $\qquad \square$

Exercise 3.15. Assume that $n \in \mathbb{N}$ is such that λ is not a spectral value of $\lambda I - P_n K$. Then verify

$$(\lambda I - P_n K)^{-1} P_n = P_n (\lambda I - K P_n)^{-1},$$

$$I + K P_n (\lambda I - K P_n)^{-1} = I + K(\lambda I - P_n K)^{-1} P_n.$$

Also, deduce (3.55) from the last equality.

Theorem 3.18. *Let N be a stability index of $(\lambda I - P_n K)$. Then the following hold for all $n \geq N$.*

(i) $\|x - x_n^S\| \leq \frac{1}{|\lambda|} \|K(x - x_n^G)\|$ *for all $n \geq N$.*

(ii) $c_1 \|K(x_n^S - x_n^G)\| \leq \|x - x_n^S\| \leq c_2 \|K(x_n^S - x_n^G)\|$.

If, in addition, $(\|P_n\|)$ is bounded, then

(iii) $c_3 \|K(I - P_n)x\| \leq \|x - x_n^S\| \leq c_4 \|K(I - P_n)x\|$,

(iv) $\|x - x_n^S\| \leq c \|K(I - P_n)\| \; \|x - x_n^G\|$.

Proof. The result in (i) follows from (3.49). Recall that x_n^S satisfies equation (3.52). Now, taking $y_n = y$ and $K_n = K P_n$ in (3.10) we obtain (ii) since the quantity β_n in (3.10) takes the form

$$\beta_n = \|K(I - P_n)x_n^S\| = \|K(x_n^S - x_n^G)\|.$$

Now, assume that $(\|P_n\|)$ is bounded. Hence, by Lemma 3.3, (iii) and (iv) follow from (3.53) and (3.54) respectively. $\qquad \square$

Considering the iterated form of x_n^S, namely,

$$\tilde{x}_n^S = \frac{1}{\lambda}(y + K x_n^S),$$

we see that

$$K(x_n^S - x_n^G) = \lambda(\tilde{x}_n^S - x_n^S).$$

Hence, the relation (ii) in Theorem 3.18 takes the form

$$c_1 |\lambda| \|\tilde{x}_n^S - x_n^S\| \leq \|x - x_n^S\| \leq c_2 |\lambda| \|\tilde{x}_n^S - x_n^S\|. \tag{3.56}$$

In fact, the above relation also follows from Theorem 3.15.

Remark 3.10. Theorem 3.18 shows that iterated Galerkin method is as good as Galerkin method, and if

$$\|K\| < |\lambda|,$$

then it gives less error than the Galerkin method. Also, part (iv) in Theorem 3.18 shows that the iterated Galerkin method is, in fact, a superconvergent method whenever

$$\|K(I - P_n)\| \to 0 \quad \text{as} \quad n \to \infty.$$

If X is a Hilbert space, K is a compact operator and (P_n) is a sequence of orthogonal projections which converges pointwise to I, then we do have the convergence $\|K(I - P_n)\| \to 0$ as $n \to \infty$ since, in this case, we have

$$\|K(I - P_n)\| = \|(I - P_n)K^*\|$$

as the adjoint operator K^* is also a compact operator.

We shall see in Section 3.8.3 that $\|K(I - P_n)\| \to 0$ if K has certain smoothness properties.

We have already seen an example of a sequence of projections, namely, interpolatory projections, with

$$\|K(I - P_n)\| \geq \|K\|.$$

Therefore, in a general situation we are not in a position to say whether iterated Galerkin method is superconvergent. $\Diamond$

(iii) Kantorovich method

In Galerkin and iterated Galerkin methods, we saw that the convergence rates depend on the 'unknown' solution x and that the convergence is guaranteed under the additional assumption $P_n y \to y$. In fact, Galerkin approximation converges if and only if $P_n y \to y$. We can get rid of these shortcomings if we allow y to vary over the range of K. Thus, if $y \in R(K)$, then

$$x := \frac{1}{\lambda}(y + Kx) = Kw$$

for some $w \in X$. In this case the estimate for $\|x - x_n^G\|$ in Theorem 3.17 (i) is

$$\|x - x_n^G\| \leq c_2 \|w\| \, \|(I - P_n)K\|.$$

Note that the rate of convergence, $\varepsilon_n := \|(I - P_n)K\|$, does not depend on the particular choice of y.

Now we consider a method which yields an estimate of the above form without the additional assumption $y \in R(K)$. For this, first we note that, by applying K to both sides of equation (3.30), we have

$$\lambda u - Ku = Ky \qquad (3.57)$$

with $u = Kx$. Conversely if u is the solution of (3.57), then

$$x = \frac{1}{\lambda}(y + u) \qquad (3.58)$$

is the solution of (3.30). Now the relation (3.58) motivates us to define an approximation of x by

$$x_n^K = \frac{1}{\lambda}(y + u_n^G), \qquad (3.59)$$

where (u_n^G) is the Galerkin approximation of the solution u of equation (3.57), that is,

$$\lambda u_n^G - P_n K u_n^G = P_n K y. \qquad (3.60)$$

We call x_n^K or sequence (x_n^K) as the **Kantorovich approximation** of x. This approximation procedure for second kind equations was first studied in detail by Schock ([67–69]).

Note that

$$\begin{aligned}
\lambda x_n^K &= y + u_n^G \\
&= y + \frac{1}{\lambda}[P_n K y + P_n K u_n^G] \\
&= y + P_n K x_n^K.
\end{aligned}$$

Thus, x_n^K satisfies the second kind equation

$$\lambda x_n^K - P_n K x_n^K = y.$$

From (3.58) and (3.59) we have

$$x - x_n^K = \frac{1}{\lambda}(u - u_n^G). \qquad (3.61)$$

As in (3.45) we have

$$u - u_n^G = \lambda(\lambda I - P_n K)^{-1}(I - P_n)u.$$

Thus,

$$x - x_n^K = (\lambda I - P_n K)^{-1}(I - P_n)Kx. \qquad (3.62)$$

Theorem 3.19. *Let $N \in \mathbb{N}$ be a stability index of $(\lambda I - P_n K)$. Then the following hold for all $n \geq N$.*

(i) $c_1\|(I - P_n)Kx\| \leq \|x - x_n^K\| \leq c_2\|(I - P_n)Kx\|$.
In particular,

$$\|x - x_N^K\| \leq c_2\|x\| \|(I - P_n)K\|$$

and $\|x - x_n^K\| \to 0$ *as* $n \to \infty$.

(ii) $c_3\|\tilde{x}_n^K - x_n^K\| \leq \|x - x_n^K\| \leq c_4\|\tilde{x}_n^K - x_n^K\|$.

(iii) *If* $(\|P_n\|)$ *is bounded, then*

$$\text{dist}(Kx, X_n\| \leq \|x - x_n^K\| \leq c\,\text{dist}(Kx, X_n).$$

Proof. The estimates in (i) follow from (3.62), and the convergence is a consequence of the fact that $\|(I - P_n)K\| \to 0$. The relations in (ii) follow from Theorem 3.15. As in Theorem 3.17, we have

$$dist(u, X_n) \leq \|u - u_n^G\| \leq c\,dist(u, X_n),$$

under the assumption that $(\|P_n\|)$ is a bounded sequence. Hence, we obtain (iii) by using the relation (3.61). $\qquad\square$

Remark 3.11. Theorem 3.19 (i) shows that, unlike Galerkin and iterated Galerkin approximations, the convergence of Kantorovich approximation is independent of the data y.

We have seen that in some cases, the iterated Galerkin method would give better error estimates. We shall see that the Kantorovich method can also yield better convergence rate provided the operator K has certain desirable smoothing properties. $\qquad\Diamond$

(iv) Iterated Kantorovich method

Analogous to the definition of iterated Galerkin approximation, we now define the iterated version of x_n^K, the **iterated Kantorovich approximation** as

$$\tilde{x}_n^K = \frac{1}{\lambda}\left(y + Kx_n^K\right).$$

As in (3.36) we have

$$x - \tilde{x}_n^K = \frac{1}{\lambda}K(x - x_n^K). \tag{3.63}$$

In this case we have the following theorem on error estimate. First recall from Lemma 3.3 that the sequence $(\lambda I - KP_n)$ of operators is stable.

Theorem 3.20. *Let $N \in \mathbb{N}$ be a stability index for $(\lambda I - P_n K)$. Then for all $n \geq N$,*

$$x - \tilde{x}_n^K = \frac{1}{\lambda}(\lambda I - K P_n)^{-1} K(I - P_n)(x - x_n^K).$$

In particular, if $N_1 \geq N$ is a stability index for $(\lambda I - K P_n)$, then

$$c_1 \|K(I - P_n)(x - x_n^K)\| \leq \|x - \tilde{x}_n^K\| \leq c_2 \|K(I - P_n)(x - x_n^K)\|$$

for all $n \geq N_1$.

Proof. First we note that

$$x - x_n^K = (\lambda I - P_n K)^{-1}(I - P_n)(x - x_n^K) \qquad (3.64)$$

which follows from (3.62) using the fact that

$$(I - P_n)Kx = (I - P_n)(u - u_n^G) = (I - P_n)(x - x_n^K).$$

From (3.64) we have

$$\begin{aligned}
\lambda(x - \tilde{x}_n^K) &= K(x - x_n^K) \\
&= K(\lambda I - P_n K)^{-1}(I - P_n)(x - x_n^K) \\
&= (\lambda I - K P_n)^{-1} K(I - P_n)(x - x_n^K).
\end{aligned}$$

From this, we obtain the required result. $\qquad\qquad\square$

Remark 3.12. The above theorem, in particular, shows that

$$\|x - \tilde{x}_n^K\| \leq c_2 \|K(I - P_n)\| \, \|x - x_n^K\|.$$

Thus, in case $\|K(I - P_n)\| \to 0$ as $n \to \infty$, the iterated Kantorovich approximation $\tilde{x}_n^K$ does have a better order of convergence than the Kantorovich approximation x_n^K. $\qquad\qquad\diamond$

(v) Modified projection method

Now we introduce another method based on the sequence (P_n) of projections which has some of the good properties common to iterated Galerkin method and the Kantorovich method, and also has the simplicity of the Galerkin method. This method, called **modified projection method**, considered in [43] by the author is defined by

$$x_n^M = x_n^G + \frac{1}{\lambda}(I - P_n)y. \qquad (3.65)$$

Note that

$$P_n x_n^M = x_n^G. \qquad (3.66)$$

Thus, as in the case of iterated Galerkin approximation (see (3.51)), if $X = C[a, b]$ and P_n is an interpolatory projection based on the nodes $t_i^{(n)}$, $i = 1, \ldots, n$, x_n^M also satisfies

$$x_n^M(t_i^{(n)}) = x_n^G(t_i^{(n)}), \quad i = 1, \ldots, n.$$

Using the relation (3.66), we have

$$\begin{aligned} \lambda x_n^M &= y + (\lambda x_n^G - P_n y) \\ &= y + P_n K x_n^G \\ &= y + P_n K P_n x_n^M. \end{aligned}$$

Thus, x_n^M satisfies the second kind equation

$$\lambda x_n^M - P_n K P_n x_n^M = y.$$

Using the relations

$$P_n x_n^M = x_n^G = P_n x_n^S,$$

$$(I - P_n) x_n^M = \frac{1}{\lambda}(I - P_n)y = (I - P_n)x_n^K, \tag{3.67}$$

and the identity $x = P_n x + (I - P_n)x$, we also have

$$x - x_n^M = P_n(x - x_n^S) + (I - P_n)(x - x_n^K). \tag{3.68}$$

The following theorem gives conditions under which the method converges and also provides error estimates in terms of the errors involved in x_n^S and x_n^K.

Theorem 3.21. *Let N be a stability index of $(\lambda I - P_n K)$. Then the following hold.*

(i) *If $\|y - P_n y\| \to 0$, then $\|x - x_n^M\| \to 0$.*
(ii) *If $(\|P_n\|)$ is bounded, then*

$$c_1 \|x - x_n^K\| \leq \|x - x_n^M\| \leq c_2 \max\{\|x - x_n^K\|, \|x - x_n^S\|\}$$

for all $n \geq N$.

Proof. By Theorem 3.17 (ii), $x_n^G \to x$ as $n \to \infty$. Hence, (i) follows from the definition of x_n^M. Now, suppose $(\|P_n\|)$ is bounded and $n \geq N$. Then from (3.68), we have

$$\|x - x_n^M\| \leq c \max\{\|x - x_n^K\|, \|x - x_n^S\|\} \quad \forall n \geq N,$$

where $c > 0$ is such that $\|P_n\| + \|I - P_n\| \le c$ for all $n \in \mathbb{N}$. Also, from (3.61), (3.62) and (3.67),

$$
\begin{aligned}
x - x_n^K &= (\lambda I - P_n K)^{-1}(I - P_n)Kx \\
&= (\lambda I - P_n K)^{-1}(I - P_n)(u - u_n^G) \\
&= (\lambda I - P_n K)^{-1}(I - P_n)(x - x_n^K) \\
&= (\lambda I - P_n K)^{-1}(I - P_n)(x - x_n^M)
\end{aligned}
$$

so that

$$
\|x - x_n^K\| \le c_0 \|x - x_n^M\|.
$$

Thus, (ii) follows by taking $c_1 = 1/c_0$ and $c_2 = c$. □

Remark 3.13. From Theorem 3.21, we can infer the following :

(a) If the iterated Galerkin approximation (x_n^S) and the Kantorovich approximation (x_n^K) have better orders of convergence than the Galerkin approximation (x_n^G), then (x_n^M) is also better than (x_n^G).

(b) If the iterated Galerkin approximation (x_n^S) is a better approximation than the Kantorovich approximation (x_n^K), then the modified projection approximation (x_n^M) is of the same order of convergence as Kantorovich approximation, that is,

$$
\|x - x_n^S\| = O\left(\|x - x_n^K\|\right) \implies \|x - x_n^M\| \cong \|x - x_n^K\|.
$$

◊

(vi) Iterated version of the modified projection method

Recall that the iterated versions of x_n^S and x_n^K are defined by

$$
\tilde{x}_n^S := \frac{1}{\lambda}\left(y + Kx_n^S\right) \quad \text{and} \quad \tilde{x}_n^K := \frac{1}{\lambda}\left(y + K\tilde{x}_n^K\right),
$$

respectively. Similarly, we define the iterated form of the modified projection approximation x_n^M by

$$
\tilde{x}_n^M = \frac{1}{\lambda}\left(y + Kx_n^M\right).
$$

This approximation has been considered in [52]. Clearly

$$
x - \tilde{x}_n^S = \frac{1}{\lambda}K(x - x_n^S). \tag{3.69}
$$

$$
x - \tilde{x}_n^K = \frac{1}{\lambda}K(x - x_n^K). \tag{3.70}
$$

$$x - \tilde{x}_n^M = \frac{1}{\lambda} K(x - x_n^M). \tag{3.71}$$

The following theorem shows the convergence properties of the iterated modified projection approximation $\tilde{x}_n^M$ in terms of those of x_n^S, $\tilde{x}_n^K$ and $\tilde{x}_n^S$.

Theorem 3.22. *Let N be a stability index for $(\lambda I - P_n K)$, and let $(\|K P_n\|)$ be bounded. Then for every $n \geq N$,*

$$\|x - \tilde{x}_n^M\| \leq c \max\{\|x - \tilde{x}_n^S\|, \|x - \tilde{x}_n^K\|, \|K(I - P_n)(x - x_n^S)\|\},$$

where $c = 1 + c_0 + |\lambda| + 1/|\lambda|$ with $\|K P_n\| \leq c_0$ for all $n \in \mathbb{N}$.

Proof. From (3.71) and (3.68), we have

$$
\begin{aligned}
x - \tilde{x}_n^M &= \frac{1}{\lambda} K(x - x_n^M) \\
&= \frac{1}{\lambda} K\left[P_n(x - x_n^S) + (I - P_n)(x - x_n^K)\right] \\
&= \frac{1}{\lambda} K P_n(x - x_n^S) + \frac{1}{\lambda} K(I - P_n)(x - x_n^K).
\end{aligned}
$$

Now,

$$
\begin{aligned}
K P_n(x - x_n^S) &= K(x - x_n^S) - K(I - P_n)(x - x_n^S) \\
&= \lambda(x - \tilde{x}_n^S) - K(I - P_n)(x - x_n^S),
\end{aligned}
$$

and by Theorem 3.20,

$$K(I - P_n)(x - x_n^K) = \lambda(\lambda I - K P_n)(x - \tilde{x}_n^K).$$

Hence, we have

$$
\begin{aligned}
x - \tilde{x}_n^M &= (x - \tilde{x}_n^S) - \frac{1}{\lambda} K(I - P_n)(x - x_n^S) \\
&\quad + (\lambda I - K P_n)(x - \tilde{x}_n^K).
\end{aligned}
$$

From this, we obtain the required bound for $\|x - \tilde{x}_n^M\|$. $\qquad\square$

3.8.2 *Computational aspects*

Let us now consider the computations involved in approximations $x_n^G, x_n^S, x_n^K, x_n^M$ considered above when P_n is of finite rank. Suppose P_n has the representation (3.41), that is,

$$P_n u = \sum_{j=1}^{n} f_j(u) u_j, \quad u \in X,$$

where $\{u_1, \ldots, u_n\}$ is a basis of $X_n := R(P_n)$ and $f_1, \ldots, f_n$ are continuous linear functionals on X such that $f_i(u_j) = \begin{cases} 1 \text{ if } i = j \\ 0 \text{ if } i \neq j \end{cases}$. Then we know that $x_n^G, x_n^S, x_n^K, x_n^M$ have the representations

$$x_n^G = \sum_{j=1}^{n} \alpha_j u_j,$$

$$x_n^S = \frac{1}{\lambda} \Big[y + \sum_{j=1}^{n} \alpha_j K u_j \Big],$$

$$x_n^K = \frac{1}{\lambda} \Big[y + \sum_{j=1}^{n} \beta_j u_j \Big],$$

$$x_n^M = \frac{y}{\lambda} - \sum_{j=1}^{n} \Big(\alpha_j - \frac{f_j(y)}{\lambda} \Big) u_j,$$

where $(\alpha_1, \ldots, \alpha_n)$ and $(\beta_1, \ldots, \beta_n)$ are the solutions of the equations

$$\lambda \alpha_i - \sum_{j=1}^{n} a_{ij} \alpha_j = f_i(y), \quad i = 1, \ldots, n,$$

and

$$\lambda \beta_i - \sum_{j=1}^{n} a_{ij} \beta_j = f_i(Ky), \quad i = 1, \ldots, n,$$

respectively. From the above expressions for the approximations, we can infer that the additional computation required for x_n^S than x_n^G is the computation of $K u_i$, $i = 1, \ldots, n$ whereas for x_n^K one needs $f_i(Ky)$ in place of $f_i(y)$ for $i = 1, \ldots, n$. Observe that the computations involved in obtaining x_n^M and x_n^G are the same.

The additional computations required for x_n^S and x_n^K would be more apparent if we consider the special nature of the functionals f_i, $i = 1, \ldots, n$, in certain specific cases. In this regard we consider two cases, one is collocation method and the other one is the orthogonal Galerkin method for the integral equation (3.31).

(a) Collocation method

Let $X = C[a, b]$ and $K : X \to X$ be the integral operator defined by (3.32) with $k(., .) \in C([a, b] \times [a, b])$. Let P_n be an interpolatory projection associated with the nodes $t_1, \ldots t_n$, and basis functions $\{u_1, \ldots, u_n\}$, that is, $u_i \in C[a, b]$ with

$$u_i(t_j) = \begin{cases} 1 \text{ if } i = j \\ 0 \text{ if } i \neq j, \end{cases}$$

so that

$$f_i(u) = u(t_i), \quad u \in C[a, b], \quad i = 1, \ldots, n.$$

In this case, we know that for $i = 1, \ldots, n$,

$$(Ku_i)(s) = \int_a^b k(s, t)u_i(t)dt, \quad s \in [a, b],$$

and

$$f_i(Ky) = \int_a^b k(t_i, t)y(t)dt.$$

(b) Orthogonal Galerkin method

Let X be $C[a, b]$ or $L^2[a, b]$ and $K : X \to X$ be the integral operator defined by (3.32) with $k(., .) \in C([a, b] \times [a, b])$. Let (P_n) be a sequence of orthogonal projections with respect to the L_2-inner product and let $\{u_1, \ldots, u_n\}$ be an orthonormal basis of $X_n := R(P_n)$ so that

$$P_n x = \sum_{j=1}^n \langle x, u_j \rangle u_j, \quad x \in X.$$

In this case, for $i, j = 1, \ldots, n$, we have

$$f_i(u) = \langle u, u_i \rangle = \int_a^b u(t)\overline{u_i(t)}\, dt, \quad u \in C[a, b],$$

$$(Ku_i)(s) = \int_a^b k(s, t)u_i(t)dt, \quad a \leq s \leq b,$$

$$f_i(Ky) = \langle Ky, u_i \rangle = \int_a^b \int_a^b k(s, t)y(t)\overline{u_i(t)}\, dt\, ds.$$

• ### 3.8.3 *Error estimates under smoothness assumptions*

In order to see the relative merits of the four methods considered in the last subsection, we shall consider bounded linear operators from $L^p[a,b]$ into itself with some smoothing properties, and P_n as a finite rank projection onto certain piecewise polynomial space.

For each $n \in \mathbb{N}$, consider a partition

$$\Delta_n : a = t_0^{(n)} < t_1^{(n)} < \ldots < t_{n-1}^{(n)} < t_n^{(n)} = b$$

of $[a,b]$ such that

$$h_n := \max\{t_i^{(n)} - t_{i-1}^{(n)} : i = 1, \ldots, n\} \to 0 \quad \text{as} \quad n \to \infty.$$

We shall also assume that the partition Δ_n is quasi-uniform, in the sense that there exists a constant $c > 0$ such that

$$h_n \leq c \, \min\{t_i^{(n)} - t_{i-1}^{(n)} : i = 1, \ldots, n\} \quad \forall n \in \mathbb{N}.$$

For $r \in \mathbb{N}$ with $r \geq 2$, let

$$S_n^r := \{u \in L^\infty[a,b] : u|_{(t_{i-1}^{(n)}, t_i^{(n)})} \text{ is a polynomial of degree atmost } r-1\}.$$

Smoothing properties of an operator K will be inferred by knowing that the range of K or K^* is contained in certain *Sobolev space*. For $1 \leq p \leq \infty$ and $m \in \mathbb{N}_0 := \mathbb{N} \cup \{0\}$, the **Sobolev space** W_p^m is the space of functions $u \in L_p$ such that for each $k \in \{0, 1, \ldots, m\}$, the k-th distributional derivative $u^{(k)}$ belongs to L^p. Thus, $u \in W_p^m$ if and only if $u \in L_p$ and for each $k \in \{0, 1, \ldots, m\}$, there exists $v_k \in L_p$ such that

$$\langle u, w^{(k)} \rangle = (-1)^k \langle v_k, w \rangle \quad \forall w \in C^\infty[a,b].$$

Here, and in what follows, we use the notation $L_p := L^p[a,b]$ and

$$\langle u, v \rangle := \int_a^b u(t)\overline{v(t)}\,dt.$$

We may recall that W_p^m is a Banach space with respect to the norm

$$\|u\|_{m,p} := \sum_{k=0}^m \|u^{(k)}\|_p, \quad u \in W_p^m,$$

and the identity operator from W_p^m into L_p is a compact operator (cf. Kesavan [32]). It is obvious from the definition of the Sobolev norms that if $s \leq m$, then $W_p^m \subseteq W_p^s$ and

$$\|u\|_{s,p} \leq \|u\|_{m,p} \quad \forall\, u \in W_p^m.$$

In the following, c is a generic constant which may take different values at different contexts, but independent of n.

We shall make use of the following known result (cf. De Vore [14] and Demko [13]).

Proposition 3.3. *For every $u \in W_p^m$,*

$$\text{dist}_p(u, S_n^r) := \inf\{\|u - v\|_p : v \in S_n^r\} \leq c\, h_n^\mu \|u\|_{m,p},$$

where $\mu := \min\{m, r\}$ and $c > 0$ is a constant independent of n.

Corollary 3.7. *If P_n is a projection onto S_n^r such that $(\|P_n\|)$ is bounded then*

$$\|(I - P_n)u\|_p \leq c\, h_n^\mu \|u\|_{m,p} \quad \forall\, u \in W^{m,p}.$$

Proof. Let $u \in W_p^m$ and $(\|P_n\|)$ be bounded, say $\|P_n\| \leq c_0$ for all $n \in \mathbb{N}$ for some $c_0 > 0$. Then we have

$$\|(I - P_n)u\|_p = \|(I - P_n)(u - v)\|_p \leq (1 + c_0)\|u - v\|, \quad \forall\, v \in S_n^r.$$

Hence,

$$\|(I - P_n)u\|_p \leq (1 + c_0)\text{dist}_p(u, S_n^r).$$

Now, the result follows by applying Proposition 3.3. $\qquad\square$

We shall consider the error estimates for the approximations when the projection P_n onto S_n^r is **orthogonal** in the sense that

$$\langle P_n u, v \rangle = \langle u, P_n v \rangle \quad \forall\, u, v \in L_p. \tag{3.72}$$

For the case of an interpolatory projection, we refer the reader to the papers of Sloan [74], Joe [30] and Graham and Sloan [23], specifically for the approximations x_n^G and x_n^S.

As a consequence of the quasi-uniformity of the partition Δ_n, it is known that $(\|P_n\|)$ is bounded (cf. Werschulz [77], Lemma 3.1).

We obtain the following theorem on error estimates by applying Proposition 3.3 to Theorems 3.17 and 3.19.

Theorem 3.23. *Suppose x is the solution of (3.30) with $X = L_p$ and $X_n = S_n^r$. Then the following hold.*

(i) *If $x \in W^{s,p}$, then*

$$\|x - x_n^G\|_p \leq c\, h_n^\zeta \|x\|_{s,p}, \quad \zeta := \min\{s, r\}.$$

(ii) *If* $K(L_p) \subseteq W_p^m$, *then*

$$\|x - x_n^K\|_p \leq c\, h_n^\mu \|Kx\|_{m,p}, \quad \mu := \min\{m, r\}.$$

Next we consider the concept of the *adjoint* K^* of a bounded linear operator $K : L_p \to L_p$.

For $1 \leq p < \infty$, let q be such that $1 < q \leq \infty$ and $\frac{1}{p} + \frac{1}{q} = 1$ with the understanding that $q = \infty$ when $p = 1$. Then it is known that for every continuous linear functional φ on L_p, there exists a unique $v_\varphi \in L_q$ such that

$$\varphi(u) = \langle u, v_\varphi \rangle \quad \forall u \in L_p,$$

and the map $\varphi \mapsto v_\varphi$ is a surjective linear isometry from L_p' to L_q, where L_p' denotes the space of all continuous linear functionals on L_p (cf. [51]). Now, for a bounded operator $K : L_p \to L_p$ with $1 \leq p < \infty$, we denote by $K^* : L_q \to L_q$ the unique linear operator defined by the requirement

$$\langle Ku, v \rangle = \langle u, K^*v \rangle \quad u \in L_p, \ v \in L_q.$$

Indeed, for each $v \in L_p$, the function $\varphi_v : L^p \to \mathbb{K}$ defined by

$$\varphi_v(u) = \langle Ku, v \rangle, \quad u \in L_p,$$

is a continuous linear functional on L_p, so that by the relation between L_q and L_p', there exists a unique $w \in L_q$ such that

$$\langle Ku, v \rangle = \varphi_v(u) = \int_a^b u(t)\overline{w(t)}\,dt, \quad u \in L_p.$$

Then K^*v is defined to be the element $w \in L_q$. The above operator K^* is called the **adjoint** of K. It can be seen that K^* is a bounded linear operator, and K^* is compact if and only if K is compact (cf. [51], Theorem 9.12).

Example 3.8. Let $K : L_p \to L_p$ be the integral operator,

$$(Kx)(s) := \int_a^b k(s,t)x(t)\,dt, \quad x \in L_p, \ s \in [a,b],$$

with $k(\cdot, \cdot) \in C[a,b](\times[a,b])$. Then it can be seen that

$$(K^*x)(s) := \int_a^b \overline{k(t,s)}x(t)\,dt, \quad x \in L_p, \ s \in [a,b],$$

and both K and K^* are compact operators.   ◇

For specifying the smoothing properties of the operators, we consider the class $T_{p,m,\ell}$ of all bounded linear operators T from L_p into itself such that $R(T) \subseteq W_p^m$, $R(T^*) \subseteq W_q^\ell$ and relations

$$\|Tu\|_{m,p} \leq c_1 \|u\|_p \quad \forall u \in L_p, \tag{3.73}$$

and

$$\|T^*u\|_{\ell,q} \leq c_2 \|v\|_q \quad \forall u \in L_q \tag{3.74}$$

hold for some constants c_1, c_2, where $1 \leq p < \infty$ and $m, \ell \in \mathbb{N}_0$.

Note that the requirements (3.73) and (3.74) on T and T^* are nothing but the requirements that

$$T : L_p \to W_p^m \quad \text{and} \quad T^* : L_q \to W_p^\ell$$

are continuous. Since the imbeddings

$$W_p^m \hookrightarrow L^p \quad \text{and} \quad W_q^\ell \hookrightarrow L^q$$

are compact (cf. Kesavan [32]), from Theorem 2.5 we can infer the following.

Proposition 3.4. *If $T \in T_{p,m,\ell}$ with at least one of ℓ, m is nonzero, then both $T : L_p \to L_p$ and $T^* : L_q \to L_q$ are compact operators.*

Exercise 3.16. Suppose $\ell, m \in \mathbb{N}$, and $K \in T_{p,m,0} \cup T_{p,0,\ell}$. Show that both K and K^* are compact operators. Further, if $1 \notin \sigma(K)$ show the following:
  (i) If $K \in T_{p,m,0}$, then $I - K : W_p^m \to W_p^m$ is bijective.
  (ii) If $K \in T_{p,0,\ell}$, then $I - K^* : W_p^\ell \to W_p^\ell$ is bijective.

Exercise 3.17. Show that if $K \in T_{p,m,0}$, then

$$\|x - x_n^K\|_p \leq c\, h_n^\mu \|x\|_p,$$

where $\mu := \min\{m, r\}$.

Next theorem is crucial for deriving error estimates for approximation methods for (3.30) under certain smoothing properties of the operator K.

In the following, without loss of generality, we assume that $\lambda = 1$.

Theorem 3.24. *Suppose $T \in T_{p,0,\ell}$ and K is a bounded operator on L_p such that $\|(I - P_n)K\| \to 0$ as $n \to \infty$. Assume further that T commutes with K and $1 \notin \sigma(K)$. For $u \in L_p$, let $v \in L_q$ and $u_n \in S_n^r$ be such that*

$$(I - K)u = v \quad \text{and} \quad (I - P_n K)u_n = P_n v.$$

Then

$$\|T(u - u_n)\|_p \leq c\, h_n^\rho \|u - u_n\|_p,$$

where $\rho := \min\{\ell, r\}$.

Proof.   By the duality of L_p and L_q, we have

$$\|T(u - u_n)\|_p = \sup_{w \neq 0} \frac{|\langle T(u - u_n), w \rangle|}{\|w\|_q}.$$

Hence, it is enough to show that

$$|\langle T(u - u_n), w \rangle| \leq c \, h_n^\lambda \|u - u_n\|_p \|w\|_q \quad \forall w \in L_q.$$

So, let $w \in L_q$. Since $1 \notin \sigma(K)$, it follows that $1 \notin \sigma(K^*)$, so that there exists a unique $z \in L_q$ such that $(I - K^*)z = T^* w$. Hence,

$$
\begin{aligned}
|\langle T(u - u_n), w \rangle| &= |\langle u - u_n, T^* w \rangle| \\
&= |\langle u - u_n, (I - K^*)z \rangle| \\
&= |\langle (I - K)(u - u_n), z \rangle|.
\end{aligned}
$$

But, since $(I - P_n K)(u - u_n) = (I - P_n)u$, we have

$$P_n(I - K)(u - u_n) = (I - P_n K)(u - u_n) - (I - P_n)(u - u_n) = 0$$

so that

$$(I - K)(u - u_n) = (I - P_n)(I - K)(u - u_n).$$

Thus,

$$
\begin{aligned}
|\langle T(u - u_n), w \rangle| &= |\langle (I - K)(u - u_n), z \rangle| \\
&= |\langle (I - K)(u - u_n), (I - P_n)z \rangle|.
\end{aligned}
$$

Now, since T commutes with K, T^* also commutes with K^*. Hence

$$z = (I - K^*)^{-1} T^* w = T^* (I - K^*)^{-1} w \in W_q^\ell.$$

Therefore, by Proposition 3.3,

$$
\begin{aligned}
|\langle T(u - u_n), w \rangle| &= |\langle (I - K)(u - u_n), (I - P_n)z \rangle| \\
&\leq (1 + \|K\|) \|u - u_n\|_p \|(I - P_n)z\|_q \\
&\leq c \, h_n^\rho \|u - u_n\|_p \|z\|_{\ell, p},
\end{aligned}
$$

where $\rho := \min\{\ell, r\}$. But, since $T \in \mathcal{T}_{p, 0, \ell}$, we have

$$
\begin{aligned}
\|z\|_{\ell, q} &= \|T^* (I - K^*)^{-1} w\|_{\ell, q} \\
&\leq c \, \|(I - K^*)^{-1} w\|_q \\
&\leq c' \, \|(I - K^*)^{-1}\| \, \|w\|_q.
\end{aligned}
$$

Hence,

$$|\langle T(u - u_n), w \rangle| \leq c \, h_n^\rho \|u - u_n\|_p \|w\|_q.$$

This completes the proof. $\qquad\qquad\qquad\qquad\qquad\qquad\qquad\qquad\quad \square$

Corollary 3.8. *Suppose x is the solution of (3.30). Then we have the following.*

(i) *If $K \in \mathcal{T}_{p,0,\ell}$, then*

$$\|x - x_n^S\|_p \leq c\, h_n^\rho \|x - x_n^G\|_p,$$
$$\|x - \tilde{x}_n^K\|_p \leq c\, h_n^\rho \|x - x_n^K\|_p, \quad \rho := \min\{\ell, r\}.$$

(ii) *If $K^2 \in \mathcal{T}_{p,0,\ell'}$, then*

$$\|x - \tilde{x}_n^S\|_p \leq c\, h_n^{\rho'} \|x - x_n^G\|_p, \quad \rho' := \min\{\ell', r\}.$$

Proof. Let $x \in L_p$ be the solution of (3.30). Then (i) and (ii) follow from Theorem 3.24 by observing the relations

$$x - x_n^S = K(x - x_n^G),$$
$$x - \tilde{x}_n^K = K(x - x_n^K) = K(u - u_n^G),$$
$$x - \tilde{x}_n^S = K(x - x_n^S) = K^2(x - x_n^G),$$

where u and u_n^G are as in (3.57) and (3.60), respectively. □

Corollary 3.8 together with Theorem 3.21 and Theorem 3.23 lead to the following.

Theorem 3.25. *Suppose x is the solution of (3.30). Then we have the following.*

(i) *If $x \in W_p^s$ and $K \in \mathcal{T}_{p,0,\ell}$, then*

$$\|x - x_n^S\|_p \leq c\, h_n^{\rho+\zeta} \|x\|_{s,p},$$

where $\rho := \min\{\ell, r\}$ and $\zeta := \min\{s, r\}$.

(ii) *If $x \in W_p^s$, $K(L_p) \subseteq W_p^m$ and $K \in \mathcal{T}_{p,0,\ell}$, then*

$$\|x - x_n^M\|_p \leq c \max\{h_n^\mu, \ h_n^{\rho+\zeta}\},$$

where $\mu := \min\{m, r\}$.

(iii) *If $K(L_p) \subseteq W_p^m$ and $K \in \mathcal{T}_{p,0,\ell}$, then*

$$\|x - \tilde{x}_n^K\|_p \leq c\, h_n^{\rho+\mu} \|Kx\|_{m,p}.$$

(iv) *If $x \in W_p^s$ and $K^2 \in \mathcal{T}_{p,0,t}$, then*

$$\|x - \tilde{x}_n^S\|_p \leq c\, h_n^{t'+\zeta} \|x\|_{s,p},$$

where $t' := \min\{t', r\}$.

Exercise 3.18. Write details of the proofs of Corollary 3.8 and Theorem 3.25.

Remark 3.14. It is apparent from Theorems 3.23 and 3.25 that if $x \in W_p^s$ and $K \in \mathcal{T}_{p,m,0} \cap \mathcal{T}_{p,0,\ell}$ with $s = 1$ and $r = \ell = m = 2$, then

$$\|x - x_n^G\|_p = O(h_n),$$
$$\|x - x_n^S\|_p = O(h_n^3),$$
$$\|x - x_n^K\|_p = O(h_n^2),$$
$$\|x - x_n^M\|_p = O(h_n^2).$$

Thus, x_n^M does provide better rate of convergence than x_n^G, though the computations involved in x_n^M is of the same order as in x_n^G.   ◇

Next we derive estimates for the errors $\|x - \tilde{x}_n^S\|$ and $\|x - \tilde{x}_n^M\|$ by assuming smoothness conditions on K^2. For this, we require the following theorem, which can be derived from Theorem 3.24, by observing that

$$K(I - P_n)u = (I - KP_n)K(u - u_n^G),$$

$$u - u_n^G = (I - P_n K)^{-1}(I - P_n)u.$$

However, we shall give below an independent proof as well.

Theorem 3.26. *Suppose $K \in \mathcal{T}_{p,0,\ell}$. Then*

$$\|K(I - P_n)u\|_p \le c\, h_n^\rho \|u\|_p, \quad \forall u \in L_p,$$

where $\rho := \min\{\ell, r\}$. In particular,

$$\|K(I - P_n)\|_{L_p \to L_p} \le c\, h_n^\rho, \quad \rho := \min\{\ell, r\}.$$

Proof.   Let $u \in L_p$. By the duality of L_p and L_q, we have

$$\|K(I - P_n)u\|_p = \sup_{w \ne 0} \frac{|\langle K(I - P_n)u, w \rangle|}{\|w\|_q}.$$

Hence, it is enough to show that

$$|\langle K(I - P_n)u, w \rangle| \le c\, h_n^\rho \|u\|_p \|w\|_q \quad \forall w \in L_q.$$

So, let $w \in L_q$. Then

$$|\langle K(I - P_n)u, w \rangle| = |\langle (I - P_n)u, K^* w \rangle|$$
$$= |\langle u, (I - P_n)K^* w \rangle|$$
$$\le \|u\|_p \|(I - P_n)K^* w\|_q.$$

Since $K^* w \in W_q^\ell$, by Proposition 3.3,

$$\|(I - P_n)K^* w\|_q \leq c\, h_n^\rho \|K^* w\|_{q,\ell},$$

where $\rho := \min\{\ell, r\}$. But, since $K \in \mathcal{T}_{p,0,\ell}$, we have

$$\|K^* w\|_{q,\ell} \leq c\|w\|_q.$$

Hence,

$$|\langle K(I - P_n)u, w \rangle| \leq c\, h_n^\rho \|u\|_p \|w\|_q.$$

This completes the proof. $\qquad\qquad\qquad\qquad\qquad\qquad\qquad\qquad\qquad\square$

Exercise 3.19. Derive Theorem 3.24 and Corollary 3.8 from Theorem 3.26.

The following Corollary is immediate from Theorems 3.25 and 3.26.

Corollary 3.9. *Suppose x is the solution of (3.30) and $K \in \mathcal{T}_{p,o,\ell}$. Then*

$$\|K(I - P_n)(x - x_n^S)\|_p \leq c\, h_n^\rho \|x - x_n^S\|_p,$$

where $\rho := \min\{\ell, r\}$. If, in addition, $x \in W_p^s$, then

$$\|K(I - P_n)(x - x_n^S)\|_p \leq c\, h_n^{2\rho+\zeta} \|x\|_{s,p},$$

where $\zeta := \min\{s, r\}$.

Applying the estimates obtained in Theorem 3.25(iii) & (iv) and Corollary 3.9 to Theorem 3.22 we obtain the following.

Theorem 3.27. *Suppose x is the solution of (3.30) with $x \in W_p^s$, $K \in \mathcal{T}_{p,0,\ell} \cap \mathcal{T}_{p,m,0}$ and $K^2 \in \mathcal{T}_{p,0,\ell'}$. Then*

$$\|x - \tilde{x}_n^M\|_p \leq c \max\{h_n^{\rho'+\zeta},\ h_n^{2\rho+\zeta},\ h_n^{\rho+\mu}\},$$

where $\zeta := \min\{s, r\}$, $\rho := \min\{\ell, r\}$, $\rho' := \min\{\ell', r\}$ and $\mu := \min\{m, r\}$.

Remark 3.15. Under the assumptions in Theorem 3.27, we obtain the following.

(i) If $s = 1$ and $r = \ell = m = 2$ and $\ell' = 2$, then

$$\|x - \tilde{x}_n^K\|_p = O(h_n^4),$$
$$\|x - \tilde{x}_n^S\|_p = O(h_n^3),$$
$$\|x - \tilde{x}_n^M\|_p = O(h_n^3).$$

(ii) If $s = 2$ and $\ell = m = 2$ and $r = \ell' = 3$, then

$$\|x - \tilde{x}_n^K\|_p = O(h_n^4),$$
$$\|x - \tilde{x}_n^S\|_p = O(h_n^5),$$
$$\|x - \tilde{x}_n^M\|_p = O(h_n^4).$$

Example 3.9. Consider the integral operator K defined by

$$(Ku)(s) = \int_0^1 \frac{u(t)}{1 + (s-t)^2}\, dt, \quad 0 \leq s \leq 1,$$

for $u \in L^2[0,1]$, and let P_n be the orthogonal projection onto the space X_n of all continuous piecewise linear functions on the equally spaced partition of $[0,1]$. As in Sloan [72], let $y(t) = \sqrt{t}$, $0 \leq t \leq 1$. In this case, we have

$$x(t) = \sqrt{t} + z(t), \quad 0 \leq t \leq 1,$$

for some $z \in C^\infty[0,1]$. Hence, $\mathrm{dist}(x, X_n) \leq c\, \mathrm{dist}(y, X_n)$. It is known that $\mathrm{dist}(y, X_n) \leq c\, h_n$ (cf. De Vore [14], Theorem 4.1). Thus,

$$\mathrm{dist}(x, X_n) \leq c\, h_n.$$

Since $Kx \in C^\infty[0,1]$ and K is self-adjoint, we have

$$\mathrm{dist}(Kx, X_n) \leq c\, h_n^2$$

and

$$\|K(I - P_n)\| = \|(I - P_n)K\| = O(h_n^2).$$

Thus, we have

$$\|x - x_n^G\| = O(h_n),$$
$$\|x - x_n^K\| = O(h_n^2),$$
$$\|x - x_n^S\| = O(h_n^3),$$

and hence, by Theorem 3.21 (ii),

$$\|x - x_n^M\| = O(\max\{\|x - x_n^S\|, \|x - x_n^K\|\}) = O(h^2).$$

We observe that the orders of convergence of x_n^K and x_n^S are better than the order of convergence of x_n^G, and consequently, x_n^M has a better order of convergence than x_n^G. ◇

Example 3.10. Consider the integral operator K defined by

$$(Kx)(s) = \int_0^{2\pi} \log\left|\left(\frac{s-t}{2}\right)\right| x(s)ds.$$

We observe that with $X = L^2[0,1]$, $K^* = K$. From the results in Sloan and Thomeè [73] it can be deduced that

$$K \in T_{2,m,0}, \quad K \in T_{2,0,\ell}, \quad K^2 \in T_{2,0,\ell'}$$

with $m = 1 = \ell$, $\ell' = 2$ (see also, [52]). Hence, if $y \in W_2^\nu[0, 2\pi]$, then

$$x = y + Kx \in W_2^\beta, \quad \beta := \min\{\nu, 1\},$$

and we obtain the following estimates:

$$\|x - x_n^G\|_2 = O(h_n^\beta),$$
$$\|x - x_n^K\|_2 = O(h_n),$$
$$\|x - x_n^S\|_2 = O(h_n^{1+\beta}),$$
$$\|x - \tilde{x}_n^K\|_2 = O(h_n^2),$$
$$\|x - \tilde{x}_n^S\|_2 = O(h_n^{2+\beta}),$$
$$\|x - \tilde{x}_n^M\|_2 = O(h_n^2).$$

From these estimates, relative advantage of the methods can be inferred. For instance, if $\beta < 1$, then it is possible that $\tilde{x}_n^M$ gives better accuracy than Galerkin, Kantorovich and iterated Galerkin approximations. ◊

Remark 3.16. For considering the collocation case, we consider the P_n to be an interpolatory projection with $R(P_n) = S_n^r$. In this case, the choice of the nodes of interpolation are important. One such choice is as follows: Divide the interval by partition points $t_i^{(n)}$ with $a = t_0^{(n)} < t_1^{(n)} \ldots < t_n^{(n)} = b$, and on each $[t_{i-1}^{(n)}, t_i^{(n)}]$, consider $\tau_{i,j}^{(n)}$ with $j = 1, \ldots, r$ such that

$$t_{i-1}^{(n)} \le \tau_{i,1}^{(n)} < \tau_{i,2}^{(n)} < \ldots < \tau_{i,r}^{(n)} \le t_i^{(n)}.$$

In particular, we take $\tau_{i,1}^{(n)}, \tau_{i,2}^{(n)}, \ldots, \tau_{i,r}^{(n)}$ to be points in $[t_{i-1}^{(n)}, t_i^{(n)}]$ obtained by shifting the r Gauss points $\tau_1, \ldots, \tau_r$, that is, zeroes of the rth degree Legendre polynomial on $[-1, 1]$. Thus, $\tau_{i,j}^{(n)} = f_i(\tau_j)$ for $j = 1, \ldots, r$, where

$$f_i(t) := t_{i-1}^{(n)} + \frac{t_i^{(n)} - t_{i-1}^{(n)}}{2}(t+1) \tag{3.75}$$

for $i = 1, \ldots, n$. Then the nodes of interpolatory nodes as $\tau_{i,j}^{(n)}$ with $j = 1, \ldots, r$, $i = 1, \ldots, n$. It is known that there exists $c_r > 0$ such that

$\|P_n\| \leq c_r$ for all $n \in \mathbb{N}$. Hence, Proposition 3.3 and Corollary 3.7 still hold. In this case, we have the following result proved in Sloan [74].

Theorem 3.28. *Suppose* $k(t, \cdot) \in W_1^m$ *for every* $t \in (a, b)$ *such that* $\{\|k(t, \cdot)\|_{m,1} : t \in (a, b)\}$ *is bounded and* $u \in W_1^s$ *with* $1 \leq s \leq 2r$ *and* $1 \leq m \leq r$. *Then*

$$\|K(I - P_n)u\|_\infty \leq c\, h_n^\gamma \|u\|_{s,1}, \quad \gamma := \min\{m + r, s\}.$$

The above theorem together with Corollary 3.7 will give error estimates for the approximations x_n^G, x_n^S, x_n^K, x_n^M and $\tilde{x}_n^K$. $\qquad\qquad \diamond$

3.8.4   *Quadrature methods for integral equations*

In this subsection we consider quadrature based approximation methods for the integral equation (3.31),

$$\lambda x(s) - \int_\Omega k(s,t)x(t)dt = y(s), \quad s \in \Omega, \tag{3.76}$$

where $\Omega = [a, b]$, $k(.,.) \in C(\Omega \times \Omega)$ and $x, y \in C(\Omega)$. The idea here is to replace the operator K,

$$(Ku)(s) = \int_\Omega k(s,t)u(t)dt, \quad s \in \Omega, \tag{3.77}$$

in the above equation by a finite rank operator based on a convergent quadrature formula, and thus obtaining an approximation K_n of K.

(i) Nyström and Fredholm methods

Recall from Section 3.5.3 that the Nyström approximation (K_n) of K corresponding to a convergent quadrature rule

$$\varphi_n(u) := \sum_{j=1}^{k_n} u(t_j^{(n)})w_j^{(n)}, \quad u \in C(\Omega), \tag{3.78}$$

is defined by

$$(K_n x)(s) = \sum_{j=1}^{\text{the}} k(s, t_j^{(n)})x(t_j^{(n)})w_j^{(n)}, \quad x \in C(\Omega),$$

where $t_1^{(n)}, \ldots, t_{k_n}^{(n)}$ in Ω are the *nodes* and $w_1^{(n)}, \ldots, w_{k_n}^{(n)}$ are the *weights* of the quadrature rule and (k_n) is an increasing sequence of positive integers. We have seen that the (K_n) converges pointwise to K and also it satisfies conditions (A1)–(A3) in Section 3.6. Thus, $(\lambda I - K_n)$ is stable with stability index, say N.

The **Nyström approximation** $(x_n^{\mathcal{N}})$ of the solution x of (3.76) is defined by the solutions of the equations

$$\lambda x_n^{\mathcal{N}} - K_n x_n^{\mathcal{N}} = y \qquad (3.79)$$

for $n \geq N$. We note that

$$
\begin{aligned}
x_n^{\mathcal{N}}(s) &= \frac{1}{\lambda}\Big[y(s) + (K_n x_n^{\mathcal{N}})(s)\Big] \\
&= \frac{1}{\lambda}\Big[y(s) + \sum_{j=1}^{k_n} k(s, t_j^{(n)}) x_n^{\mathcal{N}}(t_j^{(n)}) w_j^{(n)}\Big].
\end{aligned}
$$

Thus, $x_n^{\mathcal{N}}$ is determined by its values at the nodes $t_1^{(n)}, \ldots, t_{k_n}^{(n)}$. From (3.79), it also follows that

$$\lambda x_n^{\mathcal{N}}(t_i^{(n)}) - \sum_{j=1}^{k_n} k(t_i^{(n)}, t_j^{(n)}) x_n^{\mathcal{N}}(t_j^{(n)}) w_j^{(n)} = y(t_i^{(n)})$$

for $i = 1, \ldots, k_n$. Therefore, if $(\alpha_1^{(n)}, \ldots, \alpha_{k_n}^{(n)})$ is the solution of the system of equations

$$\lambda \alpha_i^{(n)} - \sum_{j=1}^{k_n} k(t_i^{(n)}, t_j^{(n)}) w_j^{(n)} \alpha_j^{(n)} = y(t_i^{(n)}), \quad i = 1, \ldots, k_n, \qquad (3.80)$$

then $x_n^{\mathcal{N}}$ is given by

$$x_n^{\mathcal{N}}(s) = \frac{1}{\lambda}\Big[y + \sum_{j=1}^{k_n} k(s, t_j^{(n)}) w_j^{(n)} \alpha_j^{(n)}\Big], \quad s \in \Omega.$$

Thus, in order to obtain the approximation $x_n^{\mathcal{N}}$, we need only to solve the system of equations (3.80).

Suppose $u_1^{(n)}, \ldots, u_{k_n}^{(n)}$ are continuous functions on Ω such that

$$u_i^{(n)}(t_j^{(n)}) = \begin{cases} 1, & j = i, \\ 0, & j \neq i \end{cases}$$

and

$$z_n = \sum_{j=1}^{k_n} \alpha_j^{(n)} u_j^{(n)}, \qquad (3.81)$$

where $(\alpha_1^{(n)}, \ldots, \alpha_n^{(n)})$ is the solution of the system (3.80). Then we see that

$$x_n^{\mathcal{N}} = \frac{1}{\lambda}(y + K_n z_n). \tag{3.82}$$

Moreover, if we consider the interpolatory projection P_n based on the nodes $t_1^{(n)}, \ldots, t_{k_n}^{(n)}$ and functions $u_1^{(n)}, \ldots, u_{k_n}^{(n)}$, that is,

$$P_n v = \sum_{j=1}^{k_n} v(t_j^{(n)}) u_j^{(n)}, \quad v \in C(\Omega), \tag{3.83}$$

then we have

$$K_n P_n = K_n \quad \text{and} \quad P_n x_n^{\mathcal{N}} = z_n, \tag{3.84}$$

so that equation (3.82) is the same as (3.79), and z_n satisfies the equation

$$\lambda z_n - P_n K_n z_n = P_n y,$$

which again is an operator equation of the second kind.

Let us observe the following about the sequence $(P_n K_n)$.

Proposition 3.5. *If (P_n) is a sequence of projections on $C[a, b]$ such that $P_n \to I$ pointwise, then $P_n K_n \to K$ pointwise and it satisfies the conditions (A1)–(A3) in Section 3.6. In particular, $(\lambda I - P_n K_n)$ is stable.*

Proof. We observe that

$$K - P_n K_n = (K - K_n) + (I - P_n) K_n.$$

Let us denote $A_n = K - K_n$ and $B_n = (I - P_n) K_n$. Since $K_n \to K$ pointwise, $P_n \to I$ pointwise, K compact and (K_n) satisfies conditions (A1)–(A3), it follows that $(\|A_n\|)$ bounded, $\|A_n K\| \to 0$ and $\|B_n\| \to 0$. Hence,

$$\|(K - P_n K_n) K\| \leq \|A_n K\| + \|B_n K\| \to 0$$

and

$$\|(K - P_n K_n)^2\| = \|A_n^2 + A_n B_n + B_n A_n + B_n^2\|$$
$$\leq \|A_n^2\| + 2\|A_n\| \, \|B_n\| + \|B_n\|^2$$
$$\to 0.$$

Thus $(P_n K_n)$ satisfies conditions (A1)–(A3). $\qquad\qquad\square$

The sequence (z_n) defined by (3.81) is called the **Fredholm approximation** of x, and we denote it by $(x_n^{\mathcal{F}})$. Thus, $x_n^{\mathcal{F}} := z_n$ and

$$\lambda x_n^{\mathcal{N}} - P_n K_n x_n^{\mathcal{F}} = P_n y.$$

Recall from (3.84) that $P_n x_n^{\mathcal{N}} = x_n^{\mathcal{F}}$. Thus, the Nyström approximation $x_n^{\mathcal{N}}$ is, in some sense, an interpolated form of the Fredholm approximation $x_n^{\mathcal{F}}$, and $x_n^{\mathcal{F}} \to x$ whenever $x_n^{\mathcal{N}} \to x$ and $P_n \to I$ pointwise.

(ii) Error estimates

Now we derive error estimates for the Nyström and Fredholm approximations.

Theorem 3.29. *Let (K_n) be the Nyström approximation of the integral operator K in (3.77) on $C[a, b]$ corresponding to a convergent quadrature rule, and let N be a stability index for $(\lambda I - K_n)$. Then*
$$c_1 \|(K - K_n)x\| \le \|x - x_n^{\mathcal{N}}\| \le c_2 \|(K - K_n)x\| \qquad (3.85)$$
for all $n \ge N$ and for some $c_1, c_2 > 0$. In particular,
$$\|x - x_n^{\mathcal{N}}\| \to 0 \quad as \quad n \to \infty.$$

Proof. From equations (3.30) and (3.79), it follows that
$$(\lambda I - K_n)(x - x_n^{\mathcal{N}}) = (K - K_n)x,$$
so that the estimates for $\|x - x_n^{\mathcal{N}}\|$ follows by the stability of $(\lambda I - K_n)$. The convergence of $(x_n^{\mathcal{N}})$ to x is a consequence of the pointwise convergence of (K_n). $\qquad \square$

Theorem 3.30. *Let (K_n) and N be as in Theorem 3.29. Let P_n be an interpolatory projection associated with the nodes of the quadrature rule. If $(\|P_n\|)$ is bounded, then*
$$\|x - x_n^{\mathcal{F}}\| \le c \max\{\|x - x_n^G\|, \ \|x - x_n^{\mathcal{N}}\|\},$$
$$\|x - x_n^{\mathcal{N}}\| \le c \max\{\|x - x_n^G\|, \ \|x - x_n^{\mathcal{F}}\|\},$$
$$\|x - x_n^G\| \le c \max\{\|x - x_n^{\mathcal{F}}\|, \ \|x - x_n^{\mathcal{N}}\|\},$$
for all $n \ge N$, where x_n^G is the Galerkin approximation associated with the projection P_n.

Proof. Suppose $(\|P_n\|)$ is bounded. Then, from the definitions of $x_n^{\mathcal{F}}$ and x_n^G, we have
$$\begin{aligned}
(\lambda I - P_n K_n)(x - x_n^{\mathcal{F}}) &= (\lambda I - P_n K_n)x - P_n y \\
&= (\lambda I - P_n K_n)x - (\lambda I - P_n K)x_n^G \\
&= (\lambda I - P_n K)(x - x_n^G) + P_n(K - K_n)x.
\end{aligned}$$
Now, boundedness of $(\|P_n\|)$ and the estimate in (3.85) imply the required estimates. $\qquad \square$

Remark 3.17. From Theorem 3.30 it is apparent that if
$$\|x - x_n^G\| \le c \|x - x_n^{\mathcal{N}}\| \quad or \quad \|x - x_n^G\| \le c \|x - x_n^{\mathcal{F}}\|,$$
then
$$\|x - x_n^{\mathcal{F}}\| \approx \|x - x_n^{\mathcal{N}}\|.$$
In fact, Theorem 3.30 shows that the order of convergence of two of the approximations x_n^G, $x_n^{\mathcal{N}}$, $x_n^{\mathcal{F}}$ are always the same. $\qquad \Diamond$

3.8.5 *Accelerated methods*

Let us again consider the second kind equation (3.30),

$$\lambda x - Kx = y \tag{3.86}$$

and its approximated version (3.33),

$$\lambda x_n - K_n x_n = y_n. \tag{3.87}$$

The method (3.87) may give certain error estimate of the form

$$\|x - x_n\| \le c\varepsilon_n,$$

for all large enough n, say for $n \ge N$, where both c and ε_n may depend on x or y, but one has $\varepsilon_n \to 0$. In order to have a better and better approximations to the exact solution, the only possible way seems to be taking larger and larger n. But, this is not feasible in applications, as the finite dimensional system to be solved becomes very large. So, in order to have better accuracy, what one would desire to have is a system for a reasonably large n, but fixed, and then obtain iterative refinements of the approximations that are obtained already. For example, one would like to have an equation of the form

$$\lambda x_N^{(k)} - K_N x_N^{(k)} = \varphi\left(y, x_N^{(k-1)}\right),$$

where the function $\varphi(\cdot, \cdot) : X \times X \to X$ must be simple enough to evaluate. There are many such procedures in the literature (see e.g. [18], [19], [12], [47]). We discuss here only one simple procedure.

We assume that the sequence $(\|K_n\|)$ is bounded and $(\lambda I - K_n)$ is stable. Observe that equation (3.86) is equivalent to

$$\lambda x - K_n x = y + (K - K_n)x. \tag{3.88}$$

This motivates the following procedure. Fix $N \in \mathbb{N}$ and let $x_N^{(0)} = 0$. For each $k \in \mathbb{N}$, define $x_N^{(k)}$ iteratively as the solution of the equation

$$\lambda x_N^{(k)} - K_N x_N^{(k)} = y + (K - K_N)x_N^{(k-1)}. \tag{3.89}$$

From equations (3.88) and (3.89) we see that

$$(\lambda I - K_N)(x - x_N^{(k)}) = (K - K_N)(x - x_N^{(k-1)}).$$

Thus, taking

$$\Delta_N = (\lambda I - K_N)^{-1}(K - K_N),$$

we have

$$x - x_N^{(k)} = \Delta_N(x - x_N^{(k-1)})$$

so that

$$x - x_N^{(k)} = \Delta_N^k x. \tag{3.90}$$

This implies

$$\|x - x_N^{(k)}\| \le \|\Delta_N\| \, \|x - x_N^{(k-1)}\|$$
$$\le \|\Delta_N\|^k \|x\|.$$

Thus, if we have norm convergence $\|K - K_n\| \to 0$ and if N is large enough so that $\|\Delta_N\| < 1$, then the error bound decreases strictly monotonically as k increases. In particular, the iterative process converges, that is,

$$x_N^{(k)} \to x \quad \text{as} \quad k \to \infty.$$

What happens if $\|\Delta_N\| \ge 1$?

In order to discuss this case of $\|\Delta_N\| \ge 1$, first we observe from (3.90) that

$$\|x - x_N^{(2k)}\| \le \|\Delta_N^2\| \, \|x - x_N^{(2k-2)}\|$$
$$\le \|\Delta_N^2\|^k \|x\|,$$
$$\|x - x_N^{(2k+1)}\| \le \|\Delta_N^2\| \, \|x - x_N^{(2k-1)}\|$$
$$\le \|\Delta_N^2\|^k \|\Delta_N x\|.$$

Thus if we have $\|\Delta_N^2\| < 1$, then and the iterative procedure converges with decrease in error bound at every alternate stage. The requirement $\|\Delta_N^2\| < 1$ will be satisfied for large enough N if (K_n) satisfies the conditions (A1)–(A3) in Section 3.6, since in that case we have

$$\Delta_N^2 = \frac{1}{\lambda}(\lambda I - K_N)^{-1}\left\{(K - K_N)^2 + (K - K_N)K_N \Delta_N\right\}.$$

We summarise the above discussion in the form of a theorem.

Theorem 3.31. *Suppose the sequence (K_n) of operators is such that there exists $N \in \mathbb{N}$ with $\lambda \notin \sigma(K_N)$. For each $k \in \mathbb{N}$, let $x_N^{(k)}$ be defined as in (3.89) with $x_N^{(0)} = 0$. Then, for each $k \in \mathbb{N}$, the inequalities*

$$\|x - x_N^{(2k)}\| \le \|\Delta_N^2\| \, \|x - x_N^{(2k-2)}\|$$
$$\le \|\Delta_N^2\|^k \|x\|,$$
$$\|x - x_N^{(2k+1)}\| \le \|\Delta_N^2\| \, \|x - x_N^{(2k-1)}\|$$
$$\le \|\Delta_N^2\|^k \|\Delta_N x\|$$

hold. In particular, if $\|\Delta_N^2\| < 1$, then $x_N^{(k)} \to x$.

Exercise 3.20. Write details of the proof of the above theorem.

3.9   Qualitative Properties of Convergence

In this section we consider some qualitative properties of convergence of
approximate solutions of well-posed equations.

3.9.1   *Uniform and arbitrarily slow convergence*

Suppose we have an approximation method to solve the well-posed equation
(3.1),

$$Tx = y, \tag{3.91}$$

which gives the approximation (x_n) of $x = T^{-1}y$. Suppose that the method
is **convergent**, i.e.,

$$\|x - x_n\| \to 0 \quad \text{as} \quad n \to \infty.$$

We may ask whether there can be an order of convergence for the method
which must hold for all $y \in Y$. In other words, the question is whether
there exists a sequence (δ_n) of positive reals with $\delta_n \to 0$ such that

$$\|x - x_n\| \le c\,\delta_n$$

for all large enough n and for all $y \in Y$. Here, the constant $c > 0$ may
depend on y. If such a sequence (δ_n) exists, then we say that the method
converges uniformly. Otherwise the method is said to **converge arbitrarily slowly** or an **arbitrarily slow convergent (ASC) method**.

Thus, a method which gives an approximation (x_n) is arbitrarily slowly
convergent if and only if for every sequence (δ_n) of positive reals with
$\delta_n \to 0$, there exists a $y \in Y$ such that the sequence $(\|x - x_n\|)$ converges
to zero more slowly than (δ_n), that is,

$$\limsup_n \frac{\|x - x_n\|}{\delta_n} = \infty.$$

We shall give a characterization, due to Schock [71], of arbitrarily slow
convergence of methods. For this purpose we consider a convergent method
which gives an approximation (x_n) to the solution x of (3.91). Corresponding to this method, consider the *reminder operators* $R_n : X \to X$ defined
by

$$R_n x = x - x_n.$$

Since we are discussing approximation methods for (3.91) using a sequence
of bounded linear operators, we assume, at the outset, that each R_n is a
bounded linear operator.

It is immediate from Theorem 2.11 that if the method is convergent, then $(\|R_n\|)$ is bounded.

Theorem 3.32. (Schock) *A convergent method with associated sequence of remainder operators R_n is arbitrarily slowly convergent if and only if*

$$\limsup_n \|R_n\| > 0.$$

Proof. Let (δ_n) be a sequence of positive real numbers such that $\delta_n \to 0$, and suppose $\limsup_n \|R_n\| > 0$. Then we obtain

$$\limsup_n \frac{\|R_n\|}{\delta_n} = \infty.$$

Therefore, by Uniform Boundedness Principle, there exists $x \in X$ such that

$$\limsup_n \frac{\|R_n x\|}{\delta_n} = \infty.$$

Thus, the method is arbitrarily slowly convergent.

Conversely, suppose that

$$\limsup_n \|R_n\| = 0.$$

Then, taking $\delta_n = \sup_{k \geq n} \|R_k\|$, we have

$$\|x - x_n\| \leq \|R_n\|\|x\| \leq \|x\|\delta_n,$$

for all $x \in X$, so that the method is uniformly convergent. $\square$

In view of our methods for the second kind equations, let us consider an approximation method for (3.91) which gives an approximation (x_n) obtained by solving the equation

$$T_n x_n = Q_n y, \qquad (3.92)$$

where (T_n) and (Q_n) are in $\mathcal{B}(X, Y)$ and $\mathcal{B}(Y)$ respectively with T_n bijective for all large enough n.

Theorem 3.33. *Suppose (T_n) and (Q_n) are in $\mathcal{B}(X, Y)$ and $\mathcal{B}(Y)$ respectively such that for every $u \in X$,*

$$\|(T_n - Q_n T)u\| \to 0 \quad as \quad n \to \infty.$$

Assume further that (T_n) is stable with stability index N. Then the method corresponding to (3.92) is convergent. The method is uniformly convergent if and only if

$$\limsup_n \|T_n - Q_n T\| = 0.$$

Proof. For $x \in X$, let $y = Tx$ and let x_n be the solution of (3.92) for $n \geq N$. Then we have

$$T_n(x - x_n) = (T_n - Q_nT)x,$$

so that

$$\frac{\|(T_n - Q_nT)x\|}{\|T_n\|} \leq \|R_nx\| \leq \|T_n^{-1}\| \|(T_n - Q_nT)x\|.$$

Let $c_1 > 0$ and $c_2 > 0$ be constants such that

$$c_1\|T_n\| \leq 1 \quad \text{and} \quad \|T_n^{-1}\| \leq c_2 \quad \forall n \geq N.$$

Then we have

$$c_1\|T_n - Q_nT\| \leq \|R_n\| \leq c_2\|T_n - Q_nT\|.$$

Thus, by Theorem 3.32, the method is uniformly convergent if and only if $\limsup_n \|T_n - Q_nT\| = 0$. $\qquad \square$

Let us apply Theorem 3.33 to equations (3.86) and (3.87) with $y_n = Q_ny$. In Theorem 3.33 we shall take $X = Y$, $T = \lambda I - K$ and $T_n = \lambda I - K_n$. In this case we have

$$T_n - Q_nT = (Q_nK - K_n) + \lambda(I - Q_n).$$

If we take $Q_n = I$, then we have that

$$\|T_n - Q_nT\| = \|K - K_n\|,$$

so that the corresponding method is uniformly convergent if and only if

$$\|K - K_n\| \to 0.$$

Example 3.11. Now, we consider some specific examples.

In the following cases (i) - (iv), let (P_n) be a uniformly bounded sequence of projections in $\mathcal{B}(X)$ such that

$$\|K - P_nK\| \to 0 \quad \text{as} \quad n \to \infty.$$

(i) If $K_n = P_nK$ and $Q_n = I$, then the corresponding method is the *Kantorovich method*, and we have

$$\|T_n - Q_nT\| = \|K - P_nK\| \to 0 \quad \text{as} \quad n \to \infty.$$

Thus, the Kantorovich method converges uniformly.

(ii) If $K_n = KP_n$ and $Q_n = I$, then the corresponding method is the *iterated Galerkin method*, and

$$\|T_n - Q_nT\| = \|K - KP_n\|.$$

Thus, the iterated Galerkin method converges uniformly if and only if

$$\|K - KP_n\| \to 0 \quad \text{as} \quad n \to \infty.$$

We have already seen cases where $\|K - KP_n\| \to 0$ and also cases in which $\|K - KP_n\| \geq \|K\|$.

(iii) If $K_n = P_nK$ and $Q_n = P_n$, then the corresponding method is the *Galerkin method*, and

$$\|T_n - Q_nT\| = |\lambda|\,\|I - P_n\|.$$

Therefore, if $P_n \neq I$ for infinitely many n, then the Galerkin method converges arbitrarily slowly.

(iv) If $K_n = P_nKP_n$ and $Q_n = I$, then the corresponding method is the *modified projection method*, and

$$\begin{aligned}
T_n - Q_nT &= K - P_nKP_n \\
&= (K - KP_n) + (I - P_n)KP_n.
\end{aligned}$$

Since $\|(I - P_n)K\| \to 0$ as $n \to \infty$ and $(\|P_n\|)$ is bounded, it follows that

$$\|T_n - Q_nT\| \to 0 \iff \|K - KP_n\| \to 0.$$

Thus, the modified projection method is uniformly convergent if and only if $\|K - KP_n\| \to 0$.

(v) Suppose (K_n) is the Nyström approximation of the integral operator K defined in (3.77). We have seen that

$$\|K - K_n\| \geq \|K\|,$$

so that the Nyström method is arbitrarily slowly convergent. $\Diamond$

3.9.2 *Modification of ASC-methods*

In this subsection, we see how an arbitrarily slowly convergent method can be modified to obtain a uniformly convergent method in the context of a second kind equation (3.86). Such issues have been dealt in [46].

In order to have a little more generality, we consider a companion problem using a bounded operator $Q : X \to X$ which commutes with K. We observe that if x is the solution of (3.86), then applying Q on both sides of (3.86), it is seen that $z := Qx$ is the solution of

$$z - Kz = Qy. \tag{3.93}$$

Conversely, if z is the solution of (3.86), then it follows that

$$x := z + (I - K)^{-1}(I - Q)y \tag{3.94}$$

is the solution of (3.86). Clearly, if $Q = I$ then (3.86) and (3.93) are the same.

Now, we consider an approximation method for solving (3.86) based on sequences (K_n) and (Q_n) of bounded operators. We assume that $(I - K_n)$ is stable with stability index N. Further conditions on (K_n) and (Q_n) will be imposed subsequently.

Note that the above assumption on K_n ensures that the equation

$$z_n - K_n z_n = Q_n y \tag{3.95}$$

is uniquely solvable for every $n \geq N$. Hence, in view of (3.94), we may define the approximation as

$$x_n := z_n + (I - K)^{-1}(I - Q)y \tag{3.96}$$

with z_n as the solution of (3.95).

In the above definition, although x_n involves the term

$$w := (I - K)^{-1}(I - Q)y,$$

that is, the problem of solving an equation of the second kind, appropriate choice of Q may avoid such problem. For example, if $Q = I$, then $w = 0$, and if $Q = K^m$ for some $m \in \mathbb{N}$, then $w = \sum_{j=0}^{m-1} K^j y$. More generally, if $p(t)$ is a polynomial with $p(1) = 0$, then we can find a polynomial $q(t)$ such that $1 - p(t) = (1 - t)q(t)$, and in that case the choice $Q := p(K)$ gives $w = q(K)y$, so that

$$x_n := z_n + q(K)y.$$

Now, we have the following result.

Theorem 3.34. *Consider the method defined by (3.95) and (3.96) with (K_n) and (Q_n) in $\mathcal{B}(X)$ such that $K_n \to K$ and $Q_n \to Q$ pointwise, and $(I - K_n)$ is stable with stability index N. Then the following hold.*

(i) *The method converges, and there exists $c > 0$ such that*

$$\|x - x_n\| \leq c \left(\|(K - K_n)Qx\| + \|(Q - Q_n)y\| \right) \quad \forall n \geq N.$$

(ii) *The method is uniformly convergent if and only if*

$$\|(K - K_n)Q + (Q - Q_n)(I - K)\| \to 0.$$

In particular, the method is uniformly convergent if

$$\|(K - K_n)Q\| \to 0 \quad and \quad \|Q - Q_n\| \to 0.$$

Proof. By the hypothesis, z_n and x_n are well defined for all $n \geq N$. Also, from (3.86), (3.93), (3.95) and (3.96), it follows that

$$
\begin{aligned}
x - x_n &= z - z_n \\
&= Qx - (I - K_n)^{-1}Q_n y \\
&= (I - K_n)^{-1}[(I - K_n)Qx - Q_n y].
\end{aligned}
$$

But, using the relation $x = y + Kx$,

$$
\begin{aligned}
(I - K_n)Qx - Q_n y &= Qx - Q_n y - K_n Qx \\
&= (Q - Q_n)y + (K - K_n)Qx.
\end{aligned}
$$

Thus,

$$x - x_n = (I - K_n)^{-1}[(K - K_n)Qx + (Q - Q_n)y]. \tag{3.97}$$

From this, we obtain (i).

To see the proof of (ii), we further observe from (3.97) that the remainder operator $R_n : X \to X$ defined by $R_n x := x - x_n$, $x \in X$, is given by

$$R_n := (I - K_n)^{-1}[(K - K_n)Q + (Q - Q_n)(I - K)], \quad n \geq N.$$

Thus,

$$(I - K_n)R_n = (K - K_n)Q + (Q - Q_n)(I - K).$$

Since

$$\frac{\|R_n\|}{\|(I - K_n)^{-1}\|} \leq \|(I - K_n)R_n\| \leq (1 + \|R_n\|)\|R\|_n$$

it follows that

$$\limsup_n \|R_n\| = 0 \iff \limsup_n \|(I - K_n)R_n\| = 0.$$

Thus, we obtain (ii). The particular case is immediate. $\square$

Remark 3.18. We have already remarked that the following specific cases of the method (3.33) are arbitrarily slowly convergent methods:

(a) Galerkin method.

(b) Iterated Galerkin method when $\|K(I - P_n)\| \not\to 0$.

(c) Modified projection method when $\|K(I - P_n)\| \not\to 0$.

(d) The Nyström method for second kind integral equations.

Now, in view of Theorem 3.34, we can modify an arbitrarily slowly convergent method to obtain a uniform convergent method as in (3.95) and (3.96) by suitable choices of Q and Q_n. For example, in the case of Nyström method one may modify it by taking $Q = K = Q_n$ or $Q = K$ and $Q_n = P_n K$ for some pointwise convergent sequence (P_n) of finite rank projections.

Also, in the case of Nyström method, if we choose

$$Q = K \quad \text{and} \quad Q_n = K_{m_n},$$

where m_n is much larger than n, then it can be easily seen that the accuracy of the method given by (3.95) and (3.96) would be much better than the ordinary Nyström method. However, for this choice, Theorem 3.34 does not guarantee uniform convergence.   ◊

Exercise 3.21. Prove the assertion in the last paragraph of Remark 3.18.

PROBLEMS

In the following X and Y denote Banach spaces.

(1) Let X be a separable Hilbert space and $\{u_n : n \in \mathbb{N}\}$ be an orthonormal basis of X. For $n \in \mathbb{N}$, let $P_n x = \sum_{i=1}^{n} \langle x, u_i \rangle u_i$, $x \in X$. Prove that P_n is an orthogonal projection for each $n \in \mathbb{N}$, and for each $x \in X$, $\|P_n x - x\| \to 0$ as $n \to \infty$.

(2) Let $T \in \mathcal{B}(X, Y)$ be bijective and $T_n \in \mathcal{B}(X, Y)$ for all $n \in \mathbb{N}$. Prove that $\|T - T_n\| \to 0$ if and only if $\|(T - T_n)T^{-1}\| \to 0$.

(3) Suppose $T \in \mathcal{B}(X, Y)$ is invertible, and (P_n) and (Q_n) are sequences of (non-identity) projections in $\mathcal{B}(X)$ and $\mathcal{B}(Y)$ respectively. Then neither $(Q_n T)$ nor $(P_n T)$ is a norm approximation of T – Why?

(4) Prove Theorem 3.4 by replacing the condition (3.15) by the convergence $\|[(T - T_n)T^{-1}]^k\| \to 0$ as $n \to \infty$ for some positive integer k.

(5) If $\mathbb{K} = \mathbb{C}$, then prove Theorem 3.4 by replacing the condition (3.15) by $r_\sigma((T - T_n)T^{-1}) \to 0$ as $n \to \infty$.

Hint: Recall spectral radius formula (2.10).

(6) Justify the following statements:

(i) A compact operator $K : X \to Y$ is invertible if and only if the spaces X and Y are finite dimensional with the same dimension.

(ii) A sequence (P_n) of projections on a finite dimensional space is point approximation of the identity operator if and only if $P_n = I$ for all large n.

(7) Show that the projection $P : X \to X$ defined above, where X is any one of $C[a, b]$ and $B[a, b]$ is a bounded linear operator with respect to the norm $\| \cdot \|_\infty$, and $\|P\| = \sum_{i=1}^{N} \|u_i\|_\infty$.

(8) Let $\Delta_n : a = t_0 < t_1^{(n)} < \ldots < t_n^{(n)} = b$ be a partition of the interval $[a, b]$. Prove that, if $\max\{t_i^{(n)} - t_{i-1}^{(n)} : i = 1, \ldots, n\} \to 0$ as $n \to \infty$, then the sequence (P_n) of interpolatory projections defined in (3.18) converges pointwise to the identity operator I on $C[a, b]$.

(9) Let $\Delta_n : a = t_0 < t_1^{(n)} < \ldots < t_n^{(n)} = b$ be a partition of the interval $[a, b]$, and for each $i \in \{1, \ldots, n\}$, let $\tau_{i1}^{(n)}, \ldots, \tau_{ir}^{(n)}$ be points in $[t_{i-1}^{(n)}, t_i^{(n)}]$ such that $t_{i-1}^{(n)} = \tau_{i1}^{(n)} < \ldots < \tau_{ir}^{(n)} = t_i^{(n)}$, and for $x \in C[t_{i-1}^{(n)}, t_i^{(n)}]$, let $L_{ir}^{(n)} x$ be the Lagrange interpolation polynomial of x based on $\tau_{i1}^{(n)}, \ldots, \tau_{ir}^{(n)}$. Define

$$(P_n x)(t) = (L_{ir}^{(n)} x)(t), \quad x \in C[a, b], \quad t_{i-1}^{(n)} \le t \le t_i^{(n)}.$$

Then show the following.

(i) P_n is a projection on $C[a, b]$ such that the restriction of $P_n x$ to the interval $[t_{i-1}^{(n)}, t_i^{(n)}]$ is a polynomial of degree less than r.

(ii) If $x \in C^r[a, b]$, then

$$\|x - P_n x\|_\infty \le c h_n \|x^{(r)}\|_\infty,$$

where $h_n := \max\{t_i^{(n)} - t_{i-1}^{(n)} : i = 1, \ldots, n\}$.

(iii) For every $x \in C[a,b]$,

$$\|x - P_n x\|_\infty \to 0 \quad \text{whenever} \quad h_n \to 0.$$

(10) Suppose φ and $\tilde{\varphi}$ are as above. Show that φ and $\tilde{\varphi}$ are continuous linear functionals on the Banach space $C[a,b]$ (with sup-norm) and

$$\|\varphi\| = b - a, \quad \|\tilde{\varphi}\| = \sum_{i=1}^{N} |w_i|.$$

(11) Suppose K is a compact operator such that its range is dense in X. Then show that (K_n) is pointwise approximation of K if and only if $(\|K_n\|)$ is bounded and $\|(K - K_n)K\| \to 0$.

(12) Prove the claims made in Remark 3.4.

(13) Give an example in which (A0) is satisfied for some nozero $\lambda \notin \sigma(K)$, but one of (A1)-(A3) is not satisfied.

(14) Give an example where (A1)–(A3) are satisfied but (K_n) is neither norm approximation nor collectively compact approximation of K.

(15) Justify: Suppose K is a compact operator with dense range and assumption (3.39) is satisfied. Then boundedness of $(\|P_n\|)$ implies the pointwise convergence of (P_n) to the identity operator.

(16) Prove that the conclusions of Theorem 3.13 hold if

$$\|(K - K_n)K_n\| \to 0 \quad \text{as} \quad n \to \infty,$$

the sequence $(\|K - K_n\|)$ is bounded and one of the following conditions is satisfied :

(i) $\|(K - K_n)K\| \to 0$.

(ii) $\|K_n(K - K_n)\| \to 0$.

(iii) $\exists N \in \mathbb{N}$ such that K_n compact for every $n \geq N$.

(iv) $\exists N \in \mathbb{N}$ such that $K - K_n$ compact for every $n \geq N$.

(17) Let (P_n) be a sequence of projections in $\mathcal{B}(X)$. For $x \in X$, $y \in X$, prove that

$$(\lambda I - KP_n)x = y \implies (\lambda I - P_n K)P_n x = P_n y.$$

(18) From Problem 17, deduce that, if $n \in \mathbb{N}$ is such that $\lambda I - P_n K$ is bijective, then the following hold:

(i) $\lambda I - KP_n$ is bijective and

$$(\lambda I - KP_n)^{-1} y = \frac{1}{\lambda}[y + K(\lambda I - P_n K)^{-1} P_n y].$$

(ii) For $y \in X$, if $x_n^G \to x := (\lambda I - K)^{-1} y$, then $x_n^S \to x$ for each $y \in X$ and if $(\|P_n\|)$ is bounded then the sequence $(\lambda I - KP_n)$ is stable.

(19) Suppose $\ell, m \in \mathbb{N}$, and $K \in \mathcal{T}_{p,m,0} \cup \mathcal{T}_{p,0,\ell}$. Show that both K and K^* are compact operators. Further, if $1 \notin \sigma(K)$ show the following:

(i) If $K \in \mathcal{T}_{p,m,0}$, then $I - K : W_p^m \to W_p^m$ is bijective.

(ii) If $K \in \mathcal{T}_{p,0,\ell}$, then $I - K^* : W_p^\ell \to W_p^\ell$ is bijective.

(20) Suppose $K \in \mathcal{T}_{p,0,\ell}$. From Theorem 3.24, deduce

$$\|K(I - P_n)u\|_p \le c\, h_n^\rho \|(I - P_n)u\|_p \quad \forall\, u \in L_p,$$

where $\rho := \min\{\ell, r\}$, and derive Theorem 3.26.

Chapter 4

Ill-Posed Equations and Their Regularizations

4.1 Ill-Posedness of Operator Equations

In this chapter we consider the operator equation

$$Tx = y, \tag{4.1}$$

when it is ill-posed. Here, $T : X \to Y$ is a linear operator between normed linear spaces X and Y. Recall that (4.1) is an *ill-posed equation* if T violates one of the following conditions for well-posedness:

- T is bijective,

- $T^{-1} : Y \to X$, if exists, is continuous.

From Lemma 2.6, it follows that (4.1) is ill-posed if and only if either T is not surjective or it is not bounded below.

Of course, if T is not surjective, then there will be $y \in Y$ such that (4.1) is not solvable. In such case, we shall look for the so-called *least residual norm solution* of minimal norm, say $\hat{x}$. Once the existence of such an $\hat{x}$ is guaranteed, then one would like to know whether the problem of finding $\hat{x}$ is well-posed. If this modified problem is also ill-posed, then one has to *regularize* the problem. That is, the original ill-posed problem is to be replaced by a family of *nearby* well-posed problems and then obtain approximations for $\hat{x}$ when the error in the data y approaches zero.

Let us illustrate the situation of ill-posedness of (4.1) when T is not bounded below: Suppose T is not bounded below. Then for each $n \in \mathbb{N}$, there exists nonzero $u_n \in X$ such that $\|Tu_n\| < (1/n)\|u_n\|$ for all $n \in \mathbb{N}$. Thus, if $x \in X$ and $x_n := x + \sqrt{n}u_n/\|u_n\|$, then

$$\|Tx_n - Tx\| = \frac{\sqrt{n}\|Tu_n\|}{\|u_n\|} < \frac{1}{\sqrt{n}} \to 0$$

but

$$\|x_n - x\| = \sqrt{n} \to \infty.$$

By Corollary 2.11, if X is a Banach space and T is a bounded operator with non-closed range, then T is not bounded below. Thus, for such operators T,

- (4.1) need not be solvable, and

- if T is injective, then the inverse operator $T^{-1} : R(T) \to X$ is not continuous.

There is a large class of operators T of practical interest for which $R(T)$ is not closed, namely, the compact operators of infinite rank (cf. Theorem 2.5). A prototype of an ill-posed compact operator equation is the Fredholm integral equation of the first kind,

$$\int_\Omega k(s,t)x(t)dm(t) = y(s), \quad s \in \Omega,$$

where $\Omega \subset \mathbb{R}^n$ and the kernel $k(.,.)$ are such that the operator K defined by

$$(Kx)(s) = \int_\Omega k(s,t)x(t)dm(t), \quad s \in \Omega,$$

is a compact operator on a suitable function space. In Section 2.1.3, we have given examples of compact integral operators.

Before going further, let us illustrate the ill-posedness of a compact operator equation and also give a specific example of an ill-posed compact operator equation of practical interest which also lead to an integral equation of the first kind.

4.1.1 *Compact operator equations*

Let X and Y be Hilbert spaces and $T : X \to Y$ be a compact operator of infinite rank. Then we know, by Theorem 2.28, that T can be represented as

$$Tx = \sum_{n=1}^{\infty} \sigma_n \langle x, u_n \rangle v_n, \quad x \in X,$$

where (σ_n) is a decreasing sequence of positive real numbers which converges to 0, and $\{u_n : n \in \mathbb{N}\}$ and $\{v_n : n \in \mathbb{N}\}$ are orthonormal bases of $N(T)^\perp$ and $\mathrm{cl}R(T)$ respectively.

Note that for $x \in X$ and $y \in Y$,

$$Tx = y \implies \langle x, u_n \rangle = \frac{\langle y, v_n \rangle}{\sigma_n} \quad \forall n \in \mathbb{N}.$$

Thus, for equation (4.1) to have a solution, it is necessary that

$$\sum_{n=1}^{\infty} \frac{|\langle y, v_n \rangle|^2}{\sigma_n^2} < \infty.$$

We also have

$$\|Tu_n\| = \sigma_n \to 0 \quad \text{as} \quad n \to \infty$$

but

$$\|u_n\| = 1 \quad \forall n \in \mathbb{N}.$$

In particular, T is not bounded below.

More generally, if $x \in X$ and $x_n := x + u_n/\sqrt{\sigma_n}$ for $n \in \mathbb{N}$, then $Tx_n = Tx + \sqrt{\sigma_n} v_n$ so that

$$\|Tx_n - Tx\| = \sqrt{\sigma_n} \to 0 \quad \text{as} \quad n \to \infty$$

but

$$\|x_n - x\| = \frac{1}{\sqrt{\sigma_n}} \to \infty \quad \text{as} \quad n \to \infty.$$

Thus, a small perturbation in the data y can result in large deviation in the solution x.

4.1.2 *A backward heat conduction problem*

Let $u(s,t)$ represent the temperature at a point s on a 'thine wire' of length ℓ at time t. Assuming that the wire is represented by the interval $[0, \ell]$ and its end-points are kept at zero temperature, it is known that $u(\cdot, \cdot)$ satisfies the partial differential equation

$$\frac{\partial u}{\partial t} = c^2 \frac{\partial^2 u}{\partial s^2} \tag{4.2}$$

with boundary condition

$$u(0,t) = 0 = u(\ell, t). \tag{4.3}$$

Here, c represents the thermal conductivity of the material that the wire made of. The problem that we discuss now is the following: Knowing the temperature at time $t = \tau$, say

$$g(s) := u(s, \tau), \quad 0 < s \leq \ell, \tag{4.4}$$

determine the temperature at $t = t_0 < \tau$, that is, find

$$f(s) := u(s, t_0), \quad 0 < s < \ell. \tag{4.5}$$

It can be seen easily that for each $n \in \mathbb{N}$,

$$u_n(s, t) := e^{-\lambda_n^2 t} \sin(\lambda_n s) \quad \text{with} \quad \lambda_n := \frac{c n \pi}{\ell}$$

satisfies (4.2) and (4.3). Let us assume for a moment that the general solution of (4.2) is of the form

$$u(s, t) := \sum_{n=1}^{\infty} a_n u_n(s, t), \quad s \in [0, \ell], \, t \geq 0.$$

Assuming that $f_0 := u(\cdot, 0)$ belongs to $L^2[0, \ell]$, we have

$$f_0(s) := u(s, 0) = \sum_{n=1}^{\infty} a_n u_n(s, 0) = \sqrt{\frac{\ell}{2}} \sum_{n=1}^{\infty} a_n \varphi_n(s), \quad s \in [0, \ell],$$

where

$$\varphi_n(s) := \sqrt{\frac{2}{\ell}} \, \sin(\lambda_n s), \quad s \in [0, \ell], \quad n \in \mathbb{N}.$$

Note that $\{\varphi_n : n \in \mathbb{N}\}$ is an orthonormal basis of $L^2[0, \ell]$. Thus, under the assumption that f_0 belongs to $L^2[0, \ell]$, we have

$$a_n = \sqrt{\frac{2}{\ell}} \, \langle f_0, \varphi_n \rangle, \quad n \in \mathbb{N},$$

so that

$$u(s, t) = \sum_{n=1}^{\infty} e^{-\lambda_n^2 t} \langle f_0, \varphi_n \rangle \varphi_n(s). \tag{4.6}$$

Since $|\varphi_n(s)| \leq \sqrt{2/\ell}$ for all $s \in [0, \ell]$, by Schwarz inequality and Theorem 2.3, we have

$$\sum_{n=1}^{\infty} |e^{-\lambda_n^2 t} \langle f_0, \varphi_n \rangle \varphi_n(s)| \leq \frac{2}{\ell} \|f_0\|_2 \Big(\sum_{n=1}^{\infty} e^{-2\lambda_n^2 t} \Big)^{1/2}$$

for every $s \in [0, \ell]$ and for every $t > 0$. Hence $u(s, t)$ is well defined for every $s \in [0, 1]$ and for every $t > 0$. Since $u(s, t)$ has to satisfy (4.4), we also must have

$$g(s) = u(s, \tau) = \sum_{n=1}^{\infty} e^{-\lambda_n^2 \tau} \langle f_0, \varphi_n \rangle \varphi_n(s). \tag{4.7}$$

Thus, we have

$$\langle g, \varphi_n \rangle = e^{-\lambda_n^2 \tau} \langle f_0, \varphi_n \rangle \quad \forall n \in \mathbb{N};$$

equivalently,

$$\langle f_0, \varphi_n \rangle = e^{\lambda_n^2 \tau} \langle g, \varphi_n \rangle \quad \forall n \in \mathbb{N}.$$

Hence, the data g must satisfy the condition

$$\sum_{n=1}^{\infty} e^{2\lambda_n^2 \tau} |\langle g, \varphi_n \rangle|^2 < \infty. \tag{4.8}$$

The above requirement can be interpreted as the additional smoothness conditions that the data g must satisfy. In particular, we can conclude that the existence of a solution f is not guaranteed under the sole assumption that the data g belongs to $L^2[0, 1]$. Again, from (4.6) we have

$$f(s) = u(s, t_0) = \sum_{n=1}^{\infty} e^{-\lambda_n^2 t_0} \langle f_0, \varphi_n \rangle \varphi_n(s), \quad s \in [0, \ell],$$

so that

$$\langle f, \varphi_n \rangle = e^{-\lambda_n^2 t_0} \langle f_0, \varphi_n \rangle \quad \forall n \in \mathbb{N},$$

and consequently, from (4.7), we obtain

$$g = \sum_{n=1}^{\infty} e^{-\lambda_n^2 (\tau - t_0)} \langle f, \varphi_n \rangle \varphi_n.$$

Thus, the problem of finding f from the knowledge of g is equivalent to that of solving the operator equation

$$Kf = g,$$

where

$$K\varphi = \sum_{n=1}^{\infty} e^{-\lambda_n^2 (\tau - t_0)} \langle \varphi, \varphi_n \rangle \varphi_n, \quad \varphi \in L^2[0, \ell]. \tag{4.9}$$

Since $e^{-\lambda_n^2 (\tau - t_0)} \to 0$ as $n \to \infty$, the above K is a compact operator from $L^2[0, \ell]$ into itself, and

$$\sigma_n := e^{-\lambda_n^2 (\tau - t_0)}, \quad n \in \mathbb{N},$$

are the singular values of K. Thus, K is not only not onto but also does not have continuous inverse from its range. In particular, the backward heat conduction problem considered above is an ill-posed problem.

We may also observe that the operator K in (4.9) has an integral representation, namely,

$$(Kf)(s) := \int_0^{\ell} k(s, \xi) f(\xi) d\xi, \quad \xi \in [0, \ell],$$

where

$$k(s, \xi) := \frac{2}{\ell} \sum_{n=1}^{\infty} e^{-\lambda_n^2 \tau} \sin(\lambda_n s) \sin(\lambda_n \xi), \quad s, \xi \in [0, \ell].$$

Thus, equation $Kf = g$ is an integral equation of the first kind with a *smooth* kernel.

Remark 4.1. For $t \geq 0$, define

$$K_t \varphi := \sum_{n=1}^{\infty} e^{-\lambda_n^2 t} \langle \varphi, \varphi_n \rangle \varphi_n, \quad \varphi \in L^2[0, \ell]. \tag{4.10}$$

Then we see that K_t is a compact, positive, self adjoint operator on $L^2[0, \ell]$ with singular values $e^{-\lambda_n^2 t}$, which are eigenvalues of K_t with corresponding eigenvectors φ_n. It can also be seen that, for positive real numbers p, t, t_1, t_2,

$$K_{t_1 + t_2} = K_{t_1} K_{t_2}, \quad K_{pt} = (K_t)^p.$$

Thus, we have $f = K_{t_0} f_0$ and $g = Kf = K_{\tau - t_0} K_{t_0} f_0 = K_\tau f_0$. ◊

Exercise 4.1. (a) Prove the assertions in Remark 4.1.

(b) For $t > 0$, let K_t be as in Remark 4.1. Then, for $0 < t_0 < \tau$, show that

$$K_{t_0} = (K_{\tau - t_0}^* K_{\tau - t_0})^\nu \quad \text{with} \quad \nu = \frac{t_0}{2(\tau - t_0)}.$$

4.2   LRN Solution and Generalized Inverse

4.2.1   *LRN solution*

Let X and Y be linear spaces and $T : X \to Y$ be a linear operator. Recall that for $y \in Y$, the operator equation (4.1):

$$Tx = y$$

has a solution if and only if $y \in R(T)$. If $y \notin R(T)$, then we look for an element $x_0 \in X$ such that Tx_0 is 'closest to' y. For talking about closeness we require Y to be endowed with a metric. We assume that Y is a normed linear space.

Suppose Y is a normed linear space and $y \in Y$. Then we may look for an $x_0 \in X$ which minimizes the function

$$x \mapsto \|Tx - y\|, \quad x \in X,$$

that is, to find $x_0 \in X$ such that
$$\|Tx_0 - y\| = \inf\{\|Tx - y\| : x \in X\}.$$
If such an x_0 exists, then we call it a **least residual norm solution** or an **LRN solution** of equation (4.1).

At this point it should be mentioned that, in literature, the terminology **least-squares solution** is used instead of *least residual norm solution*. This is because, in applications, one usually has the space Y as either $L^2[a,b]$ or $\mathbb{K}^n$ with norm $\|.\|_2$. The nomenclature *least residual norm solution* (LRN solution), used by the author elsewhere as well (see, e.g., [51]), conveys more appropriate meaning in a general setting.

Clearly, if $y \in R(T)$, then every solution of (4.1) is an LRN solution. However, if $y \notin R(T)$, then an LRN solution need not exist. Here is a simple example to illustrate this point.

Example 4.1. Let $X = C[a,b]$ with norm $\|\cdot\|_2$, $Y = L^2[a,b]$ and $T : X \to Y$ be the identity map. If we take $y \in Y$ with $y \notin X$, then using the denseness of X in Y, it follows that
$$\inf\{\|Tx - y\| : x \in X\} = \inf\{\|x - y\|_2 : x \in C[a,b]\} = 0,$$
but there is no $x_0 \in X$ such that $\|Tx_0 - y\| = \|x_0 - y\|_2 = 0.$ ◇

Exercise 4.2. Suppose X is a linear space, Y is a normed linear space, and $T : X \to Y$ is a linear operator such that $R(T)$ is dense in Y. If $R(T) \neq Y$, then show that there exists $y \in Y$ such that equation (4.1) does not have an LRN solution.

Now we give certain sufficient conditions which ensure the existence of an LRN-solution for (4.1) for every $y \in Y$.

Theorem 4.1. *Let X be a linear space, Y be a normed linear space and $T : X \to Y$ be a linear operator. If either*

(i) $R(T)$ *is finite dimensional or if*

(ii) Y *is a Hilbert space and $R(T)$ is a closed subspace of Y,*

then an LRN solution of (4.1) exists for every $y \in Y$.

Proof. Let $y \in Y$. In case (i) we apply Proposition 2.5, and in case (ii) we apply a consequence of Projection Theorem, namely Corollary 2.10. Thus, in both cases, there exists $y_0 \in R(T)$ such that
$$\|y - y_0\| = \inf\{\|y - v\| : v \in R(T)\}.$$
Now an $x_0 \in X$ such that $Tx_0 = y_0$ is an LRN solution of (4.1). □

We know that the condition that $R(T)$ closed is too strong to hold in many of the problems of practical interest. One way to relax this condition is to restrict the data y to lie in a suitable subspace of Y.

Theorem 4.2. *Let X be a linear space, Y be a Hilbert space and $T : X \to Y$ be a linear operator. Let $P : Y \to Y$ be the orthogonal projection onto $\mathrm{cl}R(T)$, the closure of $R(T)$. For $y \in Y$, the following are equivalent.*

(i) *equation (4.1) has an LRN solution.*
(ii) $y \in R(T) + R(T)^{\perp}$.
(iii) *The equation $Tx = Py$ has a solution.*

Proof. Let $y \in Y$. First we observe that, $Py \in R(T)$ if and only if $y \in R(T) + R(T)^{\perp}$. Thus, (ii) and (iii) are equivalent. Now, we show that (i) is equivalent to (iii). Note that, by Corollary 2.10, we have

$$\inf\{\|v - y\| : v \in \mathrm{cl}R(T)\} = \|Py - y\|.$$

Hence,

$$\begin{aligned}
\inf\{\|Tx - y\| : x \in X\} &= \inf\{\|v - y\| : v \in R(T)\} \\
&= \inf\{\|v - y\| : v \in \mathrm{cl}R(T)\} \\
&= \|Py - y\|.
\end{aligned}$$

Thus, equation (4.1) has an LRN solution x_0 if and only if

$$\|Tx_0 - y\| = \|Py - y\|.$$

Now, for every $x \in X$ we have

$$Tx - y = (Tx - Py) + (Py - y)$$

with

$$Tx - Py \in R(T), \quad Py - y \in N(P) = R(T)^{\perp},$$

so that

$$\|Tx - y\|^2 = \|Tx - Py\|^2 + \|Py - y\|^2.$$

Therefore, $x_0 \in X$ is an LRN solution of (4.1) if and only if $Tx_0 = Py$. This completes the proof. $\square$

Let X, Y and T be as in Theorem 4.2. For $y \in Y$, we denote the set of all LRN solutions of (4.1) by S_y, i.e.,

$$S_y := \{x \in X : \|Tx - y\| = \inf_{u \in X} \|Tu - y\|\}.$$

We know that S_y can be empty (see Example 4.1 and Exercise 4.2). By Theorem 4.2,

$$S_y \neq \varnothing \iff y \in R(T) + R(T)^{\perp}.$$

Corollary 4.1. *Let X, Y and T be as in Theorem 4.2, $y \in Y$ be such that $S_y \neq \varnothing$. If $x_0 \in S_y$, then*

$$S_y = \{x_0 + u : u \in N(T)\}.$$

In particular, if T is injective and $y \in R(T) + R(T)^{\perp}$, then (4.1) has a unique LRN solution.

Proof. It is clear that if x_0 is an LRN solution, then for every $u \in N(T)$, $x_0 + u$ is also an LRN solution. Thus, $x_0 + N(T) \subseteq S_y$. Also, if $P : Y \to Y$ is the orthogonal projection onto $\mathrm{cl}R(T)$ and $x_0 \in S_y$, then by Theorem 4.2, for every $x \in S_y$,

$$Tx = Py = Tx_0,$$

so that

$$x = x_0 + u \quad \text{with} \quad u = x - x_0 \in N(T)$$

and hence

$$S_y \subseteq \{x_0 + u : u \in N(T)\}.$$

Thus, we have proved that $S_y = x_0 + N(T)$. The particular case is immediate, since $N(T) = \{0\}$ whenever T is injective. $\square$

In general, when T is not injective, then a unique LRN solution from the set S_y of all LRN solutions is usually identified by imposing certain additional restrictions. A common procedure, suggested by applications of (4.1), is to choose an element in S_y at which certain non-negative function φ attains its infimum. The domain of φ can be a proper subset of X. The simplest case is when X itself is a normed linear space, and

$$\varphi(x) = \|x\|, \quad x \in X.$$

Another example of φ which is important in applications and also attracted the attention of many experts in the field of ill-posed problems is when X is a normed linear space and

$$\varphi(x) = \|Lx\|, \quad x \in D(L),$$

where $L : D(L) \to Z$ is a linear operator with its domain $D(L)$ is a subspace of X, and Z is another normed linear space. In applications, X, Y, Z may be certain function spaces and L may be a differential operator.

In spite of the importance of general situations, in the following, we shall be concerned mainly with the simple case of minimizing $x \mapsto \|x\|$, $x \in S_y$ when both X and Y are Hilbert spaces. The theory can also be used to deal with the case of minimizing

$$x \mapsto \|Lx\|, \quad x \in D(L) \cap S_y$$

(cf. [53]; also see Problem 5).

Theorem 4.3. *Let X and Y be Hilbert spaces, X_0 be a subspace of X and $T : X_0 \to Y$ be a linear operator with $N(T)$ closed in X. Let $y \in R(T) + R(T)^{\perp}$. Then there exists a unique $\hat{x} \in S_y$ such that*

$$\|\hat{x}\| = \inf_{x \in S_y} \|x\|.$$

In fact,

$$\hat{x} \in N(T)^{\perp} \quad and \quad \hat{x} = Qx_0,$$

where $Q : X \to X$ is the orthogonal projection onto $N(T)^{\perp}$ and x_0 is any element in S_y.

Proof.   Let $y \in R(T) + R(T)^{\perp}$. Then, by Theorem 4.2, the set S_y of all LRN solutions of (4.1) is nonempty. Let $x_0 \in S_y$. By Corollary 4.1, we have $S_y = \{x_0 + u : u \in N(T)\}$. Let $\hat{x} = Qx_0$, where $Q : X \to X$ is the orthogonal projection onto $N(T)^{\perp}$. Then, we have

$$x_0 = \hat{x} + u \quad \text{with} \quad u \in N(T).$$

This also implies that $\hat{x} = x_0 - u \in X_0$. Hence, by Theorem 4.2,

$$T\hat{x} = Tx_0 = Py,$$

where $P : Y \to Y$ is the orthogonal projection onto $\mathrm{cl}R(T)$. Thus, $\hat{x} \in S_y$. Again, by Corollary 4.1, for any $x \in S_y$, $x - \hat{x} \in N(T)$ so that by Pythagoras theorem, we have

$$\|x\|^2 = \|\hat{x}\|^2 + \|x - \hat{x}\|^2.$$

Thus,

$$\|\hat{x}\| \leq \|x\| \quad \forall\, x \in S_y.$$

Suppose there exists another element $\tilde{x} \in S_y$ such that $\|\tilde{x}\| \leq \|x\|$ for all $x \in S_y$. Then, it follows that

$$\|\tilde{x}\| \leq \hat{x}\| \leq \|\tilde{x}\|$$

so that $\|\tilde{x}\| = \|\hat{x}\|$. Using again Corollary 4.1 and Pythagoras theorem, we have $\tilde{x} - \hat{x} \in N(T)$ and

$$\|\tilde{x}\|^2 = \|\hat{x}\|^2 + \|\tilde{x} - \hat{x}\|^2.$$

Consequently, $\tilde{x} = \hat{x}$. $\square$

Corollary 4.2. *Let X, X_0, Y and T be as Theorem 4.3 and $Q : X \to X$ be the orthogonal projection onto $N(T)^\perp$. Let $y \in R(T) + R(T)^\perp$. Then*

$$\{x \in S_y : \|x\| = \inf_{u \in S_y} \|u\|\}, \quad N(T)^\perp \cap S_y \quad and \quad \{Qx : x \in S_y\}$$

are singleton sets.

Proof. By Theorem 4.3, there exists a unique $\hat{x} \in S_y$ such that

$$\{\hat{x}\} = \{x \in S_y : \|x\| = \inf_{u \in S_y} \|u\|\} = \{Qx : x \in S_y\}.$$

Thus, $\hat{x} \in N(T)^\perp \cap S_y$. Hence, it suffices to show that $N(T)^\perp \cap S_y$ does not contain more than one element. Suppose x_1, x_2 are in $N(T)^\perp \cap S_y$. Then, by Theorem 4.2, $Tx_1 = Tx_2$, so that we have

$$x_1 - x_2 \in N(T)^\perp \cap N(T) = \{0\}.$$

Thus, $x_1 = x_2$, and the proof is complete. $\qquad\square$

Remark 4.2. We may recall (Proposition 2.4) that null space of a closed operator is closed. Thus, the condition that $N(T)$ is closed in Theorem 4.3 is satisfied if $T : X_0 \to Y$ is a closed operator. $\qquad\Diamond$

4.2.2 *Generalized inverse*

Let X, Y be Hilbert spaces, X_0 be a subspace of X, and $T : X_0 \to Y$ be a linear operator with $N(T)$ closed in X. By Theorem 4.3, for every $y \in R(T) + R(T)^\perp$, the set

$$S_y := \{x \in X_0 : \|Tx - y\| = \inf_{u \in X_0} \|Tx - y\|\}$$

of LRN solutions of (4.1) is non-empty and there exists a unique $\hat{x} \in S_y$ such that

$$\|\hat{x}\| = \inf_{x \in S_y} \|x\|.$$

In fact, by Theorems 4.2 and 4.3, $\hat{x}$ is the unique element in $N(T)^\perp$ such that

$$T\hat{x} = Py,$$

where $P : Y \to Y$ is the orthogonal projection onto $\text{cl}R(T)$.

Let $T^\dagger$ be the map which associates each $y \in R(T) + R(T)^\perp$ to the unique LRN-solution of minimal norm. Then it can be seen that

$$T^\dagger : R(T) + R(T)^\perp \to X$$

is a linear operator with dense domain

$$D(T^\dagger) := R(T) + R(T)^\perp.$$

The operator $T^\dagger$ is called the **generalized inverse** or **Moore–Penrose inverse** of T, and $\hat{x} := T^\dagger y$ is called the **generalized solution** of (4.1). In fact,

$$T^\dagger y = (T|_{N(T)^\perp \cap X_0})^{-1} Py, \quad y \in D(T^\dagger).$$

Exercise 4.3. Let Y be a Hilbert space and Y_0 be a subspace of Y. Show that Y_0 is closed if and only if $Y_0 + Y_0^\perp = Y$.

Exercise 4.4. Let X, Y be Hilbert spaces, X_0 be a subspace of X, and $T : X_0 \to Y$ be a linear operator with $N(T)$ closed. Show that

(i) $D(T^\dagger)$ is dense in Y,

(ii) $T^\dagger$ is a linear operator, and

(iii) for $y \in D(T^\dagger)$, $T^\dagger y = (T|_{N(T)^\perp \cap X_0})^{-1} Py$, where $P : Y \to Y$ is the orthogonal projection onto cl$R(T)$.

Theorem 4.4. *Let X and Y be Hilbert spaces, X_0 be a subspace of X and $T : X_0 \to Y$ be a closed linear operator. Then*

(i) *$T^\dagger$ is a closed linear operator, and*

(ii) *$T^\dagger$ is continuous if and only if $R(T)$ is a closed subspace of Y.*

Proof. To prove the closedness of $T^\dagger$, let (y_n) be sequence in $D(T^\dagger)$ such that $y_n \to y$ in Y and $T^\dagger y_n \to x$ in X. Now, by the definition of $T^\dagger$ and using Theorem 4.2, we have

$$T(T^\dagger y_n) = Py_n \to Py,$$

where P is the orthogonal projection of Y onto the cl$R(T)$. Since T is a closed operator, it follows that $x \in X_0$ and $Tx = Py$. Hence, by Theorem 4.2, $y \in D(T^\dagger)$ and $x \in S_y$. Since $T^\dagger y_n \in N(T)^\perp \cap X_0$ and $N(T)^\perp$ is a closed subspace of X, we have

$$x = \lim_{n \to \infty} T^\dagger y_n \in N(T)^\perp$$

so that by Theorem 4.3, $T^\dagger y = x$. This proves that $T^\dagger$ is a closed operator. The second part of the theorem is a direct consequence of the closed graph theorem (Theorem 2.14). $\qquad\square$

In view of Theorem 4.4, the problem of finding a generalized solution for (4.1) is ill-posed if and only if $R(T)$ is not a closed subspace of Y. In this context we may recall from Theorem 2.17 that every compact operator of infinite rank between Banach spaces has non-closed range.

The following example shows that a large class of operators of practical interest having non-closed range.

Example 4.2. Let $X = Y = L^2[a, b]$. Consider the Fredholm integral operator K,

$$(Kx)(s) = \int_a^b k(s, t)x(t)dm(t), \quad a \le s \le b,$$

where $k(\cdot, \cdot) \in L^2([a, b] \times [a, b])$. We know that $K : L^2[a, b] \to L^2[a, b]$ is a compact operator (cf. Example 2.18). We claim that K is of finite rank if and only if $k(\cdot, \cdot)$ is *degenerate* in the sense that there exists a finite number of functions $\phi_1, \ldots, \phi_n$ and $\psi_1, \ldots, \psi_n$ in $L^2[a, b]$ such that

$$k(s, t) = \sum_{i=1}^n \phi_i(s)\psi_i(t), \quad s, t \in [a, b].$$

Clearly, if $k(\cdot, \cdot)$ is degenerate, then K is of finite rank. Conversely, suppose that K is of finite rank, say $R(K) = \operatorname{span}\{\phi_1, \ldots, \phi_n\}$. Then, for any $x \in L^2[a, b]$, there exist scalars $\alpha_1(x), \ldots, \alpha_n(x)$ such that

$$Kx = \sum_{i=1}^n \alpha_i(x)\phi_i.$$

It can be seen that, for each $i \in \{1, \ldots, n\}$, $\alpha_i(\cdot)$ is a continuous linear functional on $L^2[a, b]$. Hence, by Riesz representation theorem (cf. Theorem 2.20), there exist $u_1, \ldots, u_n$ in $L^2[a, b]$ such that $\alpha_i(x) = \langle x, u_i \rangle$ for every $x \in L^2[a, b]$. Thus,

$$Kx = \sum_{i=1}^n \alpha_i(x)\phi_i = \sum_{i=1}^n \langle x, u_i \rangle \phi_i \quad \forall x \in L^2[a, b]$$

so that, for every $x \in L^2[a, b]$,

$$\int_a^b k(s, t)x(t)dm(t) = \sum_{i=1}^n \phi_i(s) \int_a^b x(t)\overline{u_i(t)}\, dt$$

$$= \int_a^b \Big(\sum_{i=1}^n \phi_i(s)\overline{u_i(t)} \Big) x(t)\, dt.$$

Therefore, by taking $\psi_i(t) := \overline{u_i(t)}$, $t \in [a, b]$, $i \in \{1, \ldots, n\}$, we have

$$k(s, t) = \sum_{i=1}^n \phi_i(s)\psi_i(t), \quad s, t \in [a, b].$$

The above discussion shows that a Fredholm integral operator has non-closed range if the kernel is non-degenerate. $\Diamond$

4.2.3 Normal equation

Now, we give another characterization of an LRN-solution. Although the following theorem is given for a bounded operator between Hilbert spaces, the result can be shown to be true for a closed densely defined operator as well.

Theorem 4.5. *Let X and Y be Hilbert spaces, $T \in \mathcal{B}(X,Y)$ and $y \in D(T^\dagger)$. Then $x \in X$ is an LRN-solution of (4.1) if and only if*

$$T^*Tx = T^*y. \tag{4.11}$$

Proof. Let $P : Y \to Y$ be the orthogonal projection onto $\mathrm{cl}R(T)$. By Theorem 4.2, equation (4.1) has an LRN solution $x \in X$ if and only if $y \in D(T^\dagger)$, if and only if $Tx = Py$. Now,

$$
\begin{aligned}
Tx = Py &\iff P(Tx - y) = 0 \\
&\iff Tx - y \in R(T)^\perp = N(T^*) \\
&\iff T^*Tx = T^*y.
\end{aligned}
$$

Thus, the proof is complete. □

The equation

$$T^*Tx = T^*y$$

in Theorem 4.5 is called the **normal form** of equation (4.1).

As a corollary to Theorem 4.5 we have the following.

Proposition 4.1. *Let X and Y be Hilbert spaces, and $T \in \mathcal{B}(X,Y)$. Then $T^\dagger T : X \to X$ is the orthogonal projection onto $N(T)^\perp$. If, in addition, $R(T)$ is closed, then $TT^\dagger : Y \to Y$ is the orthogonal projection onto $R(T)$, and*

$$N(T)^\perp = R(T^*) = R(T^*T) = N(T^*T)^\perp.$$

In particular, if $R(T)$ is closed, then both $R(T^*)$ and $R(T^*T)$ are closed in X and are equal.

Proof. Let $P : Y \to Y$ be the orthogonal projection onto $\mathrm{cl}R(T)$ and $Q : X \to X$ be the orthogonal projection onto $N(T)^\perp$. We observe that for every $x \in X$, $Tx = P(Tx)$. Hence, by Theorem 4.3,

$$T^\dagger T = Q.$$

Also, by Theorem 4.2,

$$T(T^\dagger y) = Py$$

for every $y \in D(T^\dagger)$. Now suppose that $R(T)$ is closed in Y. Then by Theorem 4.4, $T^\dagger \in \mathcal{B}(Y, X)$ and $D(T^\dagger) = Y$, so that

$$TT^\dagger = P.$$

Next, assume that $R(T)$ is closed. Note that

$$R(T^*T) \subseteq R(T^*) \subseteq N(T)^\perp = N(T^*T)^\perp.$$

Therefore, the proof will be over, once we prove

$$N(T)^\perp \subseteq R(T^*T). \tag{4.12}$$

Since $T^\dagger T = Q$ and $T^\dagger \in \mathcal{B}(Y, X)$,

$$Q = Q^* = T^*(T^\dagger)^* u \in R(T^*).$$

Thus,

$$N(T)^\perp \subseteq R(T^*). \tag{4.13}$$

Also, for every $v \in R(T^*)$, there exists $y \in Y$ such that $v = T^*y$. Since $R(T)$ is closed in Y, by Theorem 4.1, there exists $x \in X$ such that $T^*Tx = T^*y$. Thus,

$$R(T^*) \subseteq R(T^*T). \tag{4.14}$$

Combining (4.13) and (4.14) we get (4.12), that completes the proof. □

Remark 4.3. Proposition 4.1 provides another proof for the known result in functional analysis, namely, if $T \in \mathcal{B}(X, Y)$, where X and Y are Hilbert spaces, then

$$R(T) \text{ closed } \iff R(T^*) \text{ closed}.$$

Here is another observation:

Suppose X and Y are Hilbert spaces, and $T \in \mathcal{B}(X, Y)$ with $R(T)$ is closed. Then by Proposition 4.1, we have $R(T^*T) = N(T^*T)^\perp$ so that the map $A : N(T^*T)^\perp \to R(T^*T)$ defined by

$$Ax = T^*Tx, \quad x \in N(T^*T)^\perp,$$

is bijective. Thus, from the normal form (4.11) of equation (4.1), it follows that

$$T^\dagger = A^{-1}T^*.$$

In particular, if $X = Y$ and T itself is invertible, then $T^\dagger = T^{-1}$.   ◊

Exercise 4.5. Let X and Y be Hilbert spaces and $T \in \mathcal{B}(X, Y)$ with $R(T)$ closed in Y. Show that $R(T^*T) = R(T^*)$ without relying on Theorem 4.1(ii).

4.2.4 *Picard criterion*

If T is a compact operator, then the generalized inverse of T can be represented by a singular system of T.

Theorem 4.6. (Picard criterion) *Let $T \in \mathcal{B}(X,Y)$ be a compact operator of infinite rank, $y \in Y$ and let $\{(\sigma_n, u_n, v_n) : n \in \mathbb{N}\}$ be a singular system for T. Then*

$$y \in D(T^\dagger) \iff \sum_{n=1}^{\infty} \frac{|\langle y, v_n \rangle|^2}{\sigma_n^2} < \infty,$$

and in that case

$$T^\dagger y = \sum_{n=1}^{\infty} \frac{\langle y, v_n \rangle}{\sigma_n} u_n.$$

Proof. Recall that

$$Tx = \sum_{n=1}^{\infty} \sigma_n \langle x, u_n \rangle v_n, \quad x \in X,$$

where $\{u_n : n \in \mathbb{N}\}$ and $\{v_n : n \in \mathbb{N}\}$ are orthonormal bases of $N(T)^\perp$ and $\mathrm{cl}R(T)$, respectively, and (σ_n) is a sequence of positive real numbers such that $\sigma_n \to 0$. Let $P : Y \to Y$ and $Q : X \to X$ be the orthogonal projection onto $\mathrm{cl}R(T)$ and $N(T)^\perp$ respectively. Then we have

$$Qx = \sum_{n=1}^{\infty} \langle x, u_n \rangle u_n \quad \text{and} \quad Py = \sum_{n=1}^{\infty} \langle y, v_n \rangle v_n$$

for every $x \in X$, $y \in Y$. Now, let $x \in X$, $y \in Y$. Using the fact that $Tu_n = \sigma_n v_n$ for every $n \in \mathbb{N}$, we have $Tx = Py$ if and only if

$$\sum_{n=1}^{\infty} \sigma_n \langle x, u_n \rangle v_n = \sum_{n=1}^{\infty} \langle y, v_n \rangle v_n,$$

if and only if $\sigma_n \langle x, u_n \rangle = \langle y, v_n \rangle$ for every $n \in \mathbb{N}$. Hence, by Theorems 4.2 and 4.3, it follows that $y \in D(T^\dagger)$ if and only if there exists $x \in X$ such that $Tx = Py$, if and only if

$$\sum_{n=1}^{\infty} \frac{|\langle y, v_n \rangle|^2}{\sigma_n^2} < \infty,$$

and in that case,

$$T^\dagger y = Qx = \sum_{n=1}^{\infty} \frac{\langle y, v_n \rangle}{\sigma_n} u_n.$$

This completes the proof. $\qquad\qquad\qquad\qquad\qquad\qquad\qquad\qquad\qquad\square$

The above theorem illustrates the discontinuity of $T^\dagger$ for a compact operator of infinite rank as follows: Taking $y_n = y + \sqrt{\sigma_n}\, v_n$ we have

$$\|y - y_n\| = \sqrt{\sigma_n} \to 0 \quad \text{as} \quad n \to \infty$$

but

$$\|T^\dagger y - T^\dagger y_n\| = \frac{1}{\sqrt{\sigma_n}} \to \infty \quad \text{as} \quad n \to \infty.$$

4.3 Regularization Procedure

Let X and Y be Hilbert spaces, and X_0 be a dense subspace of X. Let $T : X_0 \to Y$ be a closed linear operator. Recall from Theorem 4.4 that if $R(T)$ is not closed in Y, then the generalized inverse $T^\dagger$ of T is not continuous, and therefore the problem of finding the generalized solution $\hat{x} := T^\dagger y$ of (4.1) is ill-posed. In such situation we would like to obtain stable approximations to $\hat{x} := T^\dagger y$. Procedures that lead to stable approximations to an ill-posed problems are called *regularization methods*.

4.3.1 *Regularization family*

The first requirement for a **regularization method** for (4.1) is a family $\{R_\alpha\}_{\alpha > 0}$ of operators in $\mathcal{B}(Y, X)$ such that

$$R_\alpha y \to T^\dagger y \quad \text{as} \quad \alpha \to 0$$

for every $y \in D(T^\dagger)$. Such a family $\{R_\alpha\}_{\alpha > 0}$ of operators is called a **regularization family** for T, and for each $\alpha > 0$,

$$x_\alpha(y) := R_\alpha y$$

is called a **regularized solution** of the ill-posed equation (4.1).

In practice, one may have to work with an *inexact data* $\tilde{y}$ in place of y. In such situation what one would desire to have is the convergence of $(R_\alpha \tilde{y})$ to, say $\tilde{x}$, as $\alpha \to 0$, such that $\|\hat{x} - \tilde{x}\|$ is small whenever $\|y - \tilde{y}\|$ is small. Unfortunately, this is not always possible.

Proposition 4.2. *Let X and Y be Hilbert spaces, X_0 be a subspace of X and $T : X_0 \to Y$ be a closed linear operator. Then a regularization family $\{R_\alpha\}_{\alpha > 0}$ for T is uniformly bounded if and only if $R(T)$ is closed in Y.*

Proof. Suppose $\{R_\alpha\}_{\alpha>0}$ is uniformly bounded. Since $D(T^\dagger)$ is a dense subspace of Y and since $R_\alpha y \to T^\dagger y$ as $\alpha \to 0$ for every $y \in D(T^\dagger)$, by Theorem 2.8, it follows that $(R_\alpha y)$ converges as $\alpha \to 0$ for every y in Y and the limiting operator, say R_0, defined by

$$R_0 y := \lim_{\alpha \to 0} R_\alpha y, \quad y \in Y,$$

belongs to $\mathcal{B}(Y,X)$. But

$$T^\dagger y = R_0 y \quad \forall y \in D(T^\dagger)$$

so that $T^\dagger$, the restriction of the bounded operator R_0, is also a bounded operator. Therefore by Theorem 4.4, $R(T)$ is closed in Y.

Conversely suppose that $R(T)$ is closed in Y. Then $D(T^\dagger) = Y$ and therefore, by the hypothesis, $(R_\alpha y)$ converges as $\alpha \to 0$ for every $y \in Y$. Hence, Corollary 2.10 implies that the family $\{R_\alpha\}_{\alpha>0}$ is uniformly bounded. $\qquad\square$

The following corollary is a consequence of Proposition 4.2.

Corollary 4.3. *Let X, Y and T be as in Proposition 4.2 and $\{R_\alpha\}_{\alpha>0}$ be a regularization family for T. If $R(T)$ is not closed in Y, then for every $\delta > 0$ and for every $y \in D(T^\dagger)$, there exists $\tilde{y} \in Y$ such that $\|y - \tilde{y}\| \leq \delta$ and $\{\|R_\alpha \tilde{y}\|\}_{\alpha>0}$ is unbounded.*

Proof. Suppose $R(T)$ is not closed. By Proposition 4.2, $\{R_\alpha\}_{\alpha>0}$ is not uniformly bounded. Hence, by uniform boundedness principle (Theorem 2.9), there exists $v \in Y$ such that $\{\|R_\alpha v : \alpha > 0\|$ is not bounded. Now, let $y \in D(T^\dagger)$ and $\delta > 0$. Let

$$\tilde{y} := y + \frac{\delta v}{2\|v\|}.$$

Clearly, $\|y - \tilde{y}\| \leq \delta$. Also,

$$R_\alpha \tilde{y} := R_\alpha y + \frac{\delta R_\alpha v}{2\|v\|} \quad \forall \alpha > 0.$$

Since $R_\alpha y \to T^\dagger y$ as $\alpha \to 0$ and $\{\|R_\alpha v : \alpha > 0\|$ is not bounded, it follows that $\{\|R_\alpha \tilde{y} : \alpha > 0\|$ is also not bounded. $\qquad\square$

Suppose $\tilde{y}$ is an inexact data available in a neighbourhood of the exact data $y \in D(T^\dagger)$. Then the above corollary shows the importance of choosing the **regularization parameter** α depending on $\tilde{y}$ in such a way that

$$R_\alpha \tilde{y} \to T^\dagger y \quad \text{as} \quad \|y - \tilde{y}\| \to 0.$$

4.3.2 Regularization algorithm

Suppose that $\{R_\alpha\}_{\alpha>0}$ is a regularization family for T. We have observed that if we are given an inexact data $\tilde{y}$ in place of $y \in D(T^\dagger)$ and if $R(T)$ is not closed, then $\{R_\alpha \tilde{y}\}_{\alpha>0}$ need not converge unless $R(T)$ is closed in Y.

Let us assume that $y \in D(T^\dagger)$ and $\tilde{y} \in Y$ is such that

$$\|y - \tilde{y}\| \leq \delta$$

for some known error level $\delta > 0$. For the case of $R(T)$ closed, we have the following result.

Theorem 4.7. *Suppose $R(T)$ is closed. Then for every $y \in Y$,*

$$R_\alpha \tilde{y} \to T^\dagger y \quad as \quad \alpha \to 0,\, \delta \to 0.$$

In fact, for every $y \in Y$,

$$\|T^\dagger y - R_\alpha \tilde{y}\| \leq \|T^\dagger y - R_\alpha y\| + c\delta$$

where $c \geq \sup\limits_{\alpha>0} \|R_\alpha\|$.

Proof. Since $R(T)$ is closed, we have $D(T^\dagger) = Y$ so that for every $y \in Y$, $\|T^\dagger y - R_\alpha y\| \to 0$ as $\alpha \to 0$. Also, by Proposition 4.2, there exists $c > 0$ such that $\|R_\alpha\| \leq c$ for all $\alpha > 0$. Hence,

$$\|R_\alpha y - R_\alpha \tilde{y}\| \leq \|R_\alpha\| \, \|y - \tilde{y}\| \leq c\delta.$$

From this, the results follow. $\qquad\square$

Suppose $R(T)$ is not closed. Then our attempt must be to choose the regularization parameter α depending on δ or $\tilde{y}$ or on both such that

$$R_\alpha \tilde{y} \to \hat{x} \quad as \quad \delta \to 0,$$

where $\hat{x} := T^\dagger y$ for $y \in D(T^\dagger)$. Notice that

$$\hat{x} - R_\alpha \tilde{y} = (\hat{x} - R_\alpha y) + R_\alpha(y - \tilde{y}),$$

where the term $\hat{x} - R_\alpha y$ corresponds to the error in the regularization method, and the term $R_\alpha(y - \tilde{y})$ corresponds to the error in the data. Since $R_\alpha y \to \hat{x}$ as $\alpha \to 0$, the attempt should be to choose the parameter $\alpha = \alpha(\delta, \tilde{y})$ in such a way that

$$\alpha(\delta, \tilde{y}) \to 0 \quad and \quad R_{\alpha(\delta,\tilde{y})}(y - \tilde{y}) \to 0 \quad as \quad \delta \to 0.$$

A regularization family $\{R_\alpha : \alpha > 0\}$ together with a parameter choice strategy

$$(\delta, \tilde{y}) \mapsto \alpha := \alpha(\delta, \tilde{y})$$

is called a **regularization algorithm** if

$$\alpha(\delta, \tilde{y}) \to 0 \quad \text{and} \quad R_{\alpha(\delta, \tilde{y})}\tilde{y} \to T^\dagger y \quad \text{as} \quad \delta \to 0$$

for every $y \in D(T^\dagger)$.

We observe that

$$\|R_\alpha(y - \tilde{y})\| \le \|R_\alpha\|\,\delta.$$

We know that the family $(\|R_\alpha\|)_{\alpha>0}$ is unbounded whenever $R(T)$ is not closed. Thus, if we can choose $\alpha := \alpha(\delta)$ such that

$$\alpha(\delta) \to 0 \quad \text{and} \quad \|R_{\alpha(\delta)}\|\,\delta \to 0 \quad \text{as} \quad \delta \to 0,$$

then

$$R_{\alpha(\delta)}^{\bullet}\tilde{y}) \to T^\dagger y \quad \text{as} \quad \delta \to 0.$$

This procedure may also lead to an estimate for $\|T^\dagger y - R_\alpha \tilde{y}\|$. For instance, suppose we know a priorily that $\hat{x} := T^\dagger y$ belongs to certain subset M of X and there exist functions f and g on $[0, \infty)$ such that

$$\|T^\dagger y - R_\alpha y\| \le f(\alpha) \quad \text{and} \quad \|R_\alpha\| \le g(\alpha)$$

for all $\alpha > 0$. Then we have

$$\|T^\dagger y - R_\alpha \tilde{y}\| \le f(\alpha) + g(\alpha)\,\delta.$$

Further, if f and g are such that there exists $\alpha_\delta := \alpha(\delta)$ satisfying

$$f(\alpha_\delta) = g(\alpha_\delta)\delta,$$

then for such α_δ,

$$\|T^\dagger y - R_{\alpha_\delta}\tilde{y}\| \le 2\,f(\alpha_\delta).$$

If such functions f and g exist, then the question would be to see if the estimate obtained above is optimal in some sense. In the next section we shall consider a well-known regularization method, the so-called *Tikhonov regularization*, and discuss certain parameter choice strategies leading to convergence and error estimates.

4.4 Tikhonov Regularization

In Tikhonov regularization, the regularized solution

$$x_\alpha(y) := R_\alpha y$$

for $y \in Y$ and $\alpha > 0$ is defined as the unique element which minimizes the **Tikhonov functional**

$$x \mapsto \|Tx - y\|^2 + \alpha \|x\|^2, \quad x \in X_0.$$

The existence of the above minimizer is established in the following theorem.

Hereafter, we assume that X and Y are Hilbert spaces.

Theorem 4.8. *Let X_0 and T be as in Proposition 4.2. For each $y \in Y$ and $\alpha > 0$, there exists a unique $x_\alpha(y) \in X_0$ which minimises the map*

$$x \mapsto \|Tx - y\|^2 + \alpha \|x\|^2, \quad x \in X_0.$$

Moreover, for each $\alpha > 0$, the map

$$R_\alpha : y \mapsto x_\alpha(y), \quad y \in Y$$

is a bounded linear operator from Y to X.

Proof. Consider the product space $X \times Y$ with the usual inner product defined by

$$\langle (x_1, y_1), (x_2, y_2) \rangle_{X \times Y} = \langle x_1, x_2 \rangle_X + \langle y_1, y_2 \rangle_Y$$

for $(x_i, y_i) \in X \times Y$, $i = 1, 2$. It is easily seen that $X \times Y$ is a Hilbert space with respect to this inner product. For each $\alpha > 0$, consider the function $F_\alpha : X_0 \to X \times Y$ defined by

$$F_\alpha x = (\sqrt{\alpha}\, x, Tx), \quad x \in X_0.$$

Clearly, F_α is an injective linear operator. Since the graph of T is a closed subspace of $X \times Y$, it follows that F_α is a closed operator and $R(F_\alpha)$ is a closed subspace of $X \times Y$. Therefore, by Theorems 4.3 and 4.4, for every $y \in Y$, there exists a unique $x_\alpha \in X_0$ such that for every $x \in X_0$,

$$\begin{aligned}
\|Tx_\alpha - y\|^2 + \alpha \|x_\alpha\|^2 &= \|F_\alpha x_\alpha - (0, y)\|_{X \times Y} \\
&\leq \|F_\alpha x - (0, y)\|_{X \times Y} \\
&= \|Tx - y\|^2 + \alpha \|x\|^2,
\end{aligned}$$

and the generalized inverse $F_\alpha^\dagger$ of F_α is a continuous. In particular, the map $y \mapsto x_\alpha := F_\alpha^\dagger(0, y)$ is continuous. $\qquad \square$

For $\alpha > 0$, the unique minimizer $x_\alpha(y) \in X_0$ of the Tikhonov functional

$$x \mapsto \|Tx - y\|^2 + \alpha\|x\|^2, \quad x \in X_0$$

is called the **Tikhonov regularized solution** of (4.1).

Recall from Theorem 4.5 that if $T \in \mathcal{B}(X, Y)$, then $x \in X$ is a minimizer of the map $x \mapsto \|Tx - y\|^2$ if and only if $T^*Tx = T^*y$. Now, we show that the minimizer $x_\alpha(y)$ of the Tikhonov functional $x \mapsto \|Tx-y\|^2+\alpha\|x\|^2$ also satisfies an operator equation. In fact, we show that $x_\alpha(y)$ is the unique solution of a second-kind equation. First we prove the following lemma.

Lemma 4.1. *Suppose $A \in \mathcal{B}(X)$ is a positive self-adjoint operator. Then for every $\alpha > 0$, the operator $A + \alpha I$ is bijective and*

$$\|(A + \alpha I)^{-1}\| \leq \frac{1}{\alpha}, \tag{4.15}$$

$$\|(A + \alpha I)^{-1}A\| \leq 1. \tag{4.16}$$

Proof. Let $\alpha > 0$. Since A is positive,

$$\|(A + \alpha I)x\| \, \|x\| \geq \langle (A + \alpha I)x, x \rangle$$
$$= \langle Ax, x \rangle + \alpha\|x\|^2$$
$$\geq \alpha\|x\|^2$$

for every $x \in X$. Thus,

$$\|(A + \alpha I)x\| \geq \alpha\|x\| \quad \forall x \in X. \tag{4.17}$$

Hence, $A + \alpha I$ is injective with its range closed and its inverse from the range is continuous. Since $A + \alpha I$ is also self adjoint, we have (cf. Section 2.2.4.1)

$$R(A + \alpha I) = N(A + \alpha I)^\perp = \{0\}^\perp = X.$$

Thus, it is onto as well. Now, the relation (4.15) follows from the inequality (4.17) above.

Next we observe that

$$(A + \alpha I)^{-1}A = I - \alpha(A + \alpha I)^{-1}.$$

Hence, using the fact that $\langle (A + \alpha I)^{-1}x, x \rangle \geq 0$ for every $x \in X$, we have

$$\langle (A + \alpha I)^{-1}Ax, x \rangle = \langle (I - \alpha(A + \alpha I)^{-1})x, x \rangle$$
$$= \langle x, x \rangle - \alpha\langle (A + \alpha I)^{-1}x, x \rangle$$
$$\leq \langle x, x \rangle.$$

Since $(A + \alpha I)^{-1}A$ is self adjoint, we have $\|(A + \alpha I)^{-1}A\| \leq 1$ (cf. (2.9)). Thus, (4.16) is also proved. $\qquad\square$

Corollary 4.4. *Let $T \in \mathcal{B}(X, Y)$ and $\alpha > 0$. Then*

$$\|(T^*T + \alpha I)^{-1}T^*\| \leq \frac{1}{\alpha}.$$

Proof. Let $v \in X$. Then we have

$$
\begin{aligned}
\|(T^*T + \alpha I)^{-1}T^*v\|^2 &= \langle (T^*T + \alpha I)^{-1}T^*v, \, (T^*T + \alpha I)^{-1}T^*v \rangle \\
&= \langle T^*(TT^* + \alpha I)^{-1}v, \, T^*(TT^* + \alpha I)^{-1}v \rangle \\
&= \langle TT^*(TT^* + \alpha I)^{-1}v, \, (TT^* + \alpha I)^{-1}v \rangle \\
&= \langle (TT^* + \alpha I)^{-1}TT^*v, \, (TT^* + \alpha I)^{-1}v \rangle.
\end{aligned}
$$

Hence, by Lemma 4.1,

$$\|(T^*T + \alpha I)^{-1}T^*v\|^2 \leq \frac{\|v\|^2}{\alpha}.$$

From this, the result follows. $\qquad\square$

The relations (4.15) and (4.16) in Lemma 4.1 also follow from the following proposition, which also provide a better estimate than the one obtained in Corollary 4.4.

Proposition 4.3. *Suppose $A \in \mathcal{B}(X)$ is a positive self-adjoint operator. Then for every $\alpha > 0$,*

$$\sigma((A + \alpha I)^{-1}) = \left\{ \frac{1}{\lambda + \alpha} : \lambda \in \sigma(A) \right\}, \qquad (4.18)$$

$$\sigma((A + \alpha I)^{-1}A) = \left\{ \frac{\lambda}{\lambda + \alpha} : \lambda \in \sigma(A) \right\}, \qquad (4.19)$$

$$\sigma((A + \alpha I)^{-2}A) = \left\{ \frac{\lambda}{(\lambda + \alpha)^2} : \lambda \in \sigma(A) \right\}. \qquad (4.20)$$

Proof. By the first part of Lemma 4.1, the operator $A + \alpha I$ is bijective. To see (4.18), (4.19) and (4.20), we first observe that, for $\alpha > 0$ and $\lambda \geq 0$,

(i) $(A + \alpha I)^{-1} - (\lambda + \alpha)^{-1}I = (\lambda + \alpha)^{-1}(A + \alpha I)^{-1}(\lambda I - A)$,
(ii) $(A + \alpha I)^{-1}A = I - \alpha(A + \alpha I)^{-1}$,
(iii) $(A + \alpha I)^{-2}A = (A + \alpha I)^{-1} - \alpha(A + \alpha I)^{-2}$.

From (i), $\lambda \in \sigma(A)$ if and only if $(\lambda + \alpha)^{-1} \in \sigma((A + \alpha I)^{-1})$. Thus, we obtain (4.18). Now, relations (ii) and (iii) together with (4.18) imply

$$
\begin{aligned}
\sigma((A + \alpha I)^{-1}A) &= \{ 1 - \mu : \mu \in \sigma((A + \alpha I)^{-1}) \} \\
&= \left\{ 1 - \frac{\alpha}{\lambda + \alpha} : \lambda \in \sigma(A) \right\} \\
&= \left\{ \frac{\lambda}{\lambda + \alpha} : \lambda \in \sigma(A) \right\}
\end{aligned}
$$

and

$$\sigma((A + \alpha I)^{-2} A) = \{\mu - \alpha\mu^2 : \mu \in \sigma((A + \alpha I)^{-1})\}$$
$$= \left\{ \frac{1}{\lambda + \alpha} - \frac{\alpha}{(\lambda + \alpha)^2} : \lambda \in \sigma(A) \right\}$$
$$= \left\{ \frac{\lambda}{(\lambda + \alpha)^2} : \lambda \in \sigma(A) \right\}.$$

Thus, (4.19) and (4.19) are also proved. $\square$

Corollary 4.5. *Let* $T \in \mathcal{B}(X, Y)$ *and* $\alpha > 0$. *Then*

$$\|(T^*T + \alpha I)^{-1} T^*\| = \sup \left\{ \frac{\sqrt{\lambda}}{\lambda + \alpha} : \lambda \in \sigma(T^*T) \right\} \leq \frac{1}{2\sqrt{\alpha}}.$$

Proof.   Let R_α be as in (4.23). Then

$$R_\alpha^* R_\alpha = (T^*T + \alpha I)^{-2} T^*T.$$

Since $R_\alpha^* R_\alpha$ is a self adjoint operator, by Proposition 4.3, we have

$$\|R_\alpha\|^2 = \|R_\alpha^* R_\alpha\|$$
$$= r_\sigma(R_\alpha^* R_\alpha)$$
$$= \sup\{\lambda(\lambda + \alpha)^{-2} : \lambda \in \sigma(T^*T)\}.$$

Thus, using the relation $2\sqrt{\alpha\lambda}(\lambda + \alpha)^{-1} \leq 1$ for $\lambda, \alpha > 0$, we obtain

$$\|R_\alpha\| = \sup \left\{ \frac{\sqrt{\lambda}}{\lambda + \alpha} : \lambda \in \sigma(T^*T) \right\}$$
$$\leq \frac{1}{2\sqrt{\alpha}}.$$

This completes the proof. $\square$

Exercise 4.6. Derive the estimates in (4.15) and (4.16) from (4.18) and (4.19) respectively.

Suppose $T \in \mathcal{B}(X, Y)$. Since T^*T is a positive self-adjoint operator, by Lemma 4.1, the operator equation

$$(T^*T + \alpha I)x_\alpha = T^*y \tag{4.21}$$

has a unique solution for every $\alpha > 0$ and for every $y \in Y$, and

$$(T^*T + \alpha I)^{-1} \in \mathcal{B}(X) \quad \forall \alpha > 0,$$

that is, equation (4.21) is well-posed for every $\alpha > 0$.

In the following we denote

$$x_\alpha := (T^*T + \alpha I)^{-1} T^* y,$$

the unique solution of (4.21) and

$$\hat{x} := T^\dagger y$$

for $y \in D(T^\dagger)$. Then we note that

$$(T^*T + \alpha I)\hat{x} = T^* y + \alpha \hat{x},$$

so that

$$\hat{x} - x_\alpha = \alpha (T^*T + \alpha I)^{-1} \hat{x}. \tag{4.22}$$

Theorem 4.9. *Let* $T \in \mathcal{B}(X, Y)$, $y \in Y$ *and* $\alpha > 0$. *Then the solution* x_α *of equation (4.21) minimizes the function*

$$x \mapsto \|Tx - y\|^2 + \alpha \|x\|^2, \quad x \in X.$$

Proof. Let $x_\alpha = (T^*T + \alpha I)^{-1} T^* y$. Then for every $x \in X$, taking $u = x - x_\alpha$ we have

$$\|Tx - y\|^2 + \alpha \|x\|^2 = \|Tx_\alpha - y\|^2 + \alpha \|x_\alpha\|^2 + \langle (T^*T + \alpha I)u, u \rangle$$
$$\geq \|Tx_\alpha - y\|^2 + \alpha \|x_\alpha\|^2.$$

Thus x_α minimizes the function $x \mapsto \|Tx - y\|^2 + \alpha \|x\|^2$. $\square$

Hereafter, we assume that $T \in \mathcal{B}(X, Y)$.

Remark 4.4. Let $\alpha > 0$. Since the map $x \mapsto \|Tx - y\|^2 + \alpha \|x\|^2$ has a unique minimizer and the equation $(T^*T + \alpha I)x = T^* y$ has a unique solution for every $y \in Y$, Theorem 4.9 shows that, for $x \in X$,

$$\|Tx - y\|^2 + \alpha \|x\|^2 = \inf_{u \in X} \|Tu - y\|^2 + \alpha \|u\|^2$$

if and only if $(T^*T + \alpha I)x = T^* y$. $\Diamond$

Our next attempt is to show that $\{R_\alpha\}_{\alpha > 0}$ with

$$R_\alpha = (T^*T + \alpha I)^{-1} T^* \quad \forall \alpha > 0 \tag{4.23}$$

is a regularization family for (4.1). For this purpose, we shall make use of the relation

$$\text{cl} R(T^*T) = N(T^*T)^\perp = N(T)^\perp. \tag{4.24}$$

Lemma 4.2. *For every* $u \in N(T)^\perp$ *and* $v \in N(T^*)^\perp$,

$$\|\alpha (T^*T + \alpha I)^{-1} u\| \to 0, \quad \|\alpha (TT^* + \alpha I)^{-1} v\| \to 0$$

as $\alpha \to 0$.

Proof. Let $A_\alpha := \alpha(T^*T + \alpha I)^{-1}$ for $\alpha > 0$. By the relation (4.15) in Lemma 4.1, we know that $\|A_\alpha\| \leq 1$ for every $\alpha > 0$, and by (4.24), $R(T^*T)$ is a dense subspace of $N(T)^\perp$. Now, for $x \in R(T^*T)$, let $z \in X$ be such that $x = T^*Tz$. Then by the relation (4.16) in Lemma 4.1, we have

$$\begin{aligned}
\|A_\alpha x\| &= \|A_\alpha T^*Tz\| \\
&= \alpha\|(T^*T + \alpha I)^{-1}T^*Tz\| \\
&\leq \alpha\|v\|.
\end{aligned}$$

Thus, for every $x \in R(T^*T)$, $\|\alpha(T^*T + \alpha I)^{-1}x\| \to 0$ as $\alpha \to 0$. Therefore, by Theorem 2.8, $\|\alpha(T^*T + \alpha I)^{-1}u\| \to 0$ for every $u \in N(T)^\perp$.

By interchanging the roles of T and T^* in that above argument we also obtain $\|\alpha(TT^* + \alpha I)^{-1}v\| \to 0$ for every $u \in N(T^*)^\perp$. $\qquad\square$

Theorem 4.10. *For $y \in D(T^\dagger)$,*

$$\|\hat{x} - x_\alpha\| \to 0 \quad as \quad \alpha \to 0.$$

In particular, the family $\{R_\alpha\}_{\alpha>0}$ with R_α as in (4.23) is a regularization family.

Proof. Let $y \in D(T^\dagger)$. Recall from (4.22) that

$$\hat{x} - x_\alpha = \alpha(T^*T + \alpha I)^{-1}\hat{x}.$$

Hence, by Lemma 4.2, we have $\|\hat{x} - x_\alpha\| \to 0$ as $\alpha \to 0$. $\qquad\square$

Corollary 4.6. *Let $T \in \mathcal{B}(X, Y)$ and R_α as in (4.23). Then $\{R_\alpha\}_{\alpha>0}$ is uniformly bounded if and only if $R(T)$ is closed, and in that case*

$$\sup_{\alpha>0} \|R_\alpha\| = \|T^\dagger\|.$$

Proof. By Theorem 4.10, $\{R_\alpha\}_{\alpha>0}$ is a regularization family for T. Hence, by Proposition 4.2, $\{R_\alpha\}_{\alpha>0}$ is uniformly bounded if and only if $R(T)$ is closed. Now, suppose $R(T)$ is closed. Then by Proposition 4.1, $P := TT^\dagger$ is the orthogonal projection onto $R(T)$. Since $N(P) = R(T)^\perp = N(T^*)$, it then follows using the estimate (4.16) in Lemma 4.1 that

$$\begin{aligned}
\|R_\alpha y\| &= \|R_\alpha TT^\dagger\| \\
&= \|(T^*T + \alpha I)^{-1}T^*TT^\dagger\| \\
&\leq \|T^\dagger y\|
\end{aligned}$$

for all $y \in Y$ and for all $\alpha > 0$. Thus, $\|R_\alpha\| \leq \|T^\dagger\|$. We also have

$$\|T^\dagger y\| = \lim_{\alpha \to 0} \|R_\alpha y\| \leq \|y\| \sup_{\alpha>0} \|R_\alpha\|$$

for every $y \in Y$. This completes the proof. $\qquad\square$

Remark 4.5. For a more general form of Tikhonov regularization using an unbounded operator L, one may refer [36] or [53]. Problem 6 describes such a procedure. ◇

In Section 4.2.4 we have seen for a compact operator, how the ill-posedness of (4.1) related to the singular values of T. Next we show, using the family of operators $R_\alpha = (T^*T + \alpha I)^{-1}T^*$, $\alpha > 0$, that ill-posedness of the equation (4.1) is closely associated with the nature of the spectrum of T^*T in the neighbourhood of zero.

Theorem 4.11. *Let $T \in \mathcal{B}(X,Y)$. Then $R(T)$ is closed in Y if and only if zero is not an accumulation point of $\sigma(T^*T)$.*

Proof. Let R_α as in (4.23). Then by Corollary 4.5, we have

$$\|R_\alpha\| = \sup\left\{\sqrt{\lambda}(\lambda + \alpha)^{-1} : \lambda \in \sigma(T^*T)\right\}.$$

Let

$$f_\alpha(\lambda) = \sqrt{\lambda}(\lambda + \alpha)^{-1}, \quad \lambda \in \sigma(T^*T).$$

Now, suppose that zero is an accumulation point of $\sigma(T^*T)$. Then there exists a sequence (λ_n) in $\sigma(T^*T)$ such that $\lambda_n \to 0$ as $n \to \infty$, so that $f_{\lambda_n}(\lambda_n) \to \infty$ as $n \to \infty$. Note that

$$\|R_{\lambda_n}\| \geq f_{\lambda_n}(\lambda_n) = \frac{1}{2\sqrt{\lambda_n}} \quad \forall n \in \mathbb{N}.$$

Hence, $\|R_{\lambda_n}\| \to \infty$ as $n \to \infty$, so that by Corollary 4.6, $R(T)$ is not closed in Y.

To see the converse, suppose that zero is not an accumulation point of $\sigma(T^*T)$. Then there exists $\lambda_0 > 0$ such that

$$\sigma(T^*T) \subset \{0\} \cup [\lambda_0, \|T\|^2].$$

Since $f_\alpha(0) = 0$ and

$$f_\alpha(\lambda) \leq \frac{1}{\sqrt{\lambda}} \leq \frac{1}{\sqrt{\lambda_0}}, \quad 0 \neq \lambda \in \sigma(T^*T),$$

we have

$$\|R_\alpha\| = \sup\{f_\alpha(\lambda) : \lambda \in \sigma(T^*T)\} \leq \frac{1}{\sqrt{\lambda_0}}.$$

Thus, $\{\|R_\alpha\| : \alpha > 0\}$ is bounded. Hence, again by Corollary 4.6, $R(T)$ is closed in Y. □

Remark 4.6. Theorem 4.11 and some other characterizations of closedness of $R(T)$ are available in [37]. ◇

4.4.1   *Source conditions and order of convergence*

Let $T \in \mathcal{B}(X, Y)$. As in (4.23), we denote

$$R_\alpha := (T^*T + \alpha I)^{-1} T^*, \quad \alpha > 0,$$

and for $y \in D(T^\dagger)$, we denote

$$\hat{x} := T^\dagger y, \quad x_\alpha := R_\alpha y, \quad \alpha > 0.$$

In this subsection we obtain estimates for the error $\|\hat{x} - x_\alpha\|$ under certain assumptions on $\hat{x}$.

Theorem 4.12. *Suppose* $y \in D(T^\dagger)$.

(i) *If* $\hat{x} = T^*v$ *for some* $v \in Y$, *then*

$$\|\hat{x} - x_\alpha\| \le \|v\| \sqrt{\alpha}.$$

(ii) *If* $\hat{x} = T^*Tu$ *for some* $u \in X$, *then*

$$\|\hat{x} - x_\alpha\| \le \|u\| \alpha.$$

Proof.   Suppose $\hat{x} = T^*v$ for some $v \in Y$. Then from the relation (4.22) and the estimate in Corollary 4.4, we have

$$\begin{aligned}
\|\hat{x} - x_\alpha\| &= \|\alpha (T^*T + \alpha I)^{-1} T^*v\| \\
&\le \|v\| \sqrt{\alpha}.
\end{aligned}$$

Next, assume that $\hat{x} = T^*Tu$ for some $u \in X$. Then from the relation (4.22) and the estimate (4.16) in Lemma 4.1, we have

$$\|\hat{x} - x_\alpha\| = \|\alpha (T^*T + \alpha I)^{-1} T^*Tu\| \le \|u\| \alpha.$$

This completes the proof. $\square$

Corollary 4.7. *Suppose* $R(T)$ *is closed. Then for every* $y \in Y$, *there exists* $u \in X$ *such that*

$$\|\hat{x} - x_\alpha\| \le \|u\| \alpha.$$

Proof.   Since $R(T)$ is closed, by Proposition 4.1, we have

$$N(T)^\perp = R(T^*) = R(T^*T).$$

Hence, the result follows from Theorem 4.12 (ii). $\square$

In view of Theorem 4.12 (ii), A natural question would be whether it is possible to improve the order $O(\alpha)$ to $o(\alpha)$. The answer is negative except in a trivial case.

Theorem 4.13. *If* $\|\hat{x} - x_\alpha\| = o(\alpha)$, *then* $\hat{x} = 0$ *and* $x_\alpha = 0$.

Proof. From (4.22), we have

$$\|\hat{x} - x_\alpha\| = \alpha\|(T^*T + \alpha I)^{-1}\hat{x}\| \geq \frac{\alpha\|\hat{x}\|}{\alpha + \|T^*T\|}.$$

In particular,

$$\frac{\|\hat{x}\|}{\alpha + \|T^*T\|} \leq \frac{\|\hat{x} - x_\alpha\|}{\alpha}.$$

Hence,

$$\|\hat{x} - x_\alpha\| = o(\alpha) \quad \text{implies} \quad \|\hat{x}\| = 0,$$

and in that case,

$$x_\alpha = (T^*T + \alpha I)^{-1}T^*y = (T^*T + \alpha I)^{-1}T^*T\hat{x} = 0.$$

This completes the proof. $\qquad\square$

A priori assumptions on the unknown 'solution' $T^\dagger y$, as in Theorem 4.12, are called **source conditions**.

Now, let us look at the source conditions imposed on $\hat{x} := T^\dagger y$ in Theorem 4.12 in the context of T being a compact operator.

Let T be a compact operator and $\{(\sigma_n, u_n, v_n) : n \in \mathbb{N}\}$ be a singular system of T. Then, from the representations

$$Tx = \sum_{n=1}^{\infty} \sigma_n \langle x, u_n \rangle v_n, \quad x \in X, \tag{4.25}$$

$$T^*Tx = \sum_{n=1}^{\infty} \sigma_n^2 \langle x, u_n \rangle u_n, \quad x \in X, \tag{4.26}$$

we have

$$\hat{x} \in R(T^*) \iff \sum_{n=1}^{\infty} \frac{|\langle \hat{x}, u_n \rangle|^2}{\sigma_n^2} < \infty, \tag{4.27}$$

$$\hat{x} \in R(T^*T) \iff \sum_{n=1}^{\infty} \frac{|\langle \hat{x}, u_n \rangle|^2}{\sigma_n^4} < \infty. \tag{4.28}$$

Exercise 4.7. Prove (4.27) and (4.28).

We observe that the source conditions (4.27) and (4.28) are special cases of the condition

$$\sum_{n=1}^{\infty} \frac{|\langle \hat{x}, u_n \rangle|^2}{\sigma_n^{4\nu}} < \infty \tag{4.29}$$

for $\nu > 0$. The condition (4.29) is known as a **Hölder-type source condition**. We shall soon define such condition for non-compact operators as well. First let us obtain an estimate for the error $\|\hat{x} - x_\alpha\|$ under the source condition (4.29).

Estimates under Hölder-type source condition

Since $T^*Tu_n = \sigma_n^2 u_n$ for all $n \in \mathbb{N}$, we have

$$(T^*T + \alpha I)^{-1}u_n = \frac{u_n}{\sigma_n^2 + \alpha} \quad \forall \alpha > 0, \, n \in \mathbb{N}.$$

Therefore,

$$x_\alpha := R_\alpha y = (T^*T + \alpha I)^{-1}T^*y$$

$$= \sum_{n=1}^{\infty} \frac{\sigma_n}{\sigma_n^2 + \alpha} \langle y, v_n \rangle u_n.$$

Also, using the Fourier expansion $\sum_{n=1}^{\infty} \langle \hat{x}, u_n \rangle u_n$ of $\hat{x}$, we have

$$\hat{x} - x_\alpha = \alpha (T^*T + \alpha I)^{-1}\hat{x}$$

$$= \sum_{n=1}^{\infty} \frac{\alpha \langle \hat{x}, u_n \rangle}{\sigma_n^2 + \alpha} u_n.$$

Suppose $\hat{x}$ satisfies the condition (4.29) for some $\nu > 0$. Then, taking

$$u := \sum_{n=1}^{\infty} \frac{\langle \hat{x}, u_n \rangle}{\sigma_n^{2\nu}} u_n, \tag{4.30}$$

we see that $u \in N(T)^\perp$ and $\langle \hat{x}, u_n \rangle = \sigma_n^{2\nu} \langle u, u_n \rangle$ for all $n \in \mathbb{N}$ so that we have

$$\hat{x} - x_\alpha = \sum_{n=1}^{\infty} \frac{\alpha \langle \hat{x}, u_n \rangle}{\sigma_n^2 + \alpha} u_n$$

$$= \sum_{n=1}^{\infty} \frac{\alpha \sigma_n^{2\nu}}{\sigma_n^2 + \alpha} \langle u, u_n \rangle u_n.$$

Since, for $\lambda > 0$ and $0 < \nu \le 1$,

$$\frac{\lambda^\nu}{\lambda + \alpha} = \alpha^{\nu-1} \frac{(\lambda/\alpha)^\nu}{1 + \lambda/\alpha} \le \alpha^{\nu-1}$$

we obtain

$$\|\hat{x} - x_\alpha\|^2 = \sum_{n=1}^{\infty} \frac{\alpha^2 \sigma_n^{4\nu}}{(\sigma_n^2 + \alpha)^2} |\langle u, u_n \rangle|^2$$

$$\le \alpha^{2\nu} \|u\|^2.$$

Thus, we have proved the following theorem.

Theorem 4.14. *Suppose T is a compact operator with singular system $\{(\sigma_n, u_n, v_n) : n \in \mathbb{N}\}$ and $y \in D(T^\dagger)$. If $\hat{x} := T^\dagger y$ satisfies (4.29) for some $\nu \in (0, 1]$, then there exists $u \in X$ such that*

$$\|\hat{x} - x_\alpha\| \le \alpha^\nu \|u\|.$$

Remark 4.7. In view of the definition (2.14) and the representation (4.26) for the compact positive operator T^*T, we have

$$(T^*T)^\nu x = \sum_{n=1}^{\infty} \sigma_n^{2\nu} \langle x, u_n \rangle u_n, \quad x \in X, \qquad (4.31)$$

for any $\nu > 0$. Thus, the condition (4.29) on $\hat{x} := T^\dagger y$ is equivalent to the condition $\hat{x} \in R((T^*T)^\nu)$. ◇

Recall also that for every $T \in \mathcal{B}(X, Y)$, the operator T^*T is positive and self-adjoint, and hence by the discussion in Section 2.3.4, for every any $\nu > 0$, we have a positive self-adjoint operator $(T^*T)^\nu$. Now, in place of Theorem 4.14, we have the following theorem.

Theorem 4.15. *Suppose* $\hat{x} \in R((T^*T)^\nu)$ *for some* $\nu \in (0, 1]$. *Then*

$$\|\hat{x} - x_\alpha\| \le \alpha^\nu \|u\|,$$

where $u \in X$ *is such that* $\hat{x} = (T^*T)^\nu u$.

Proof. Recall that from (4.22) that

$$\hat{x} - x_\alpha = \alpha (T^*T + \alpha I)^{-1} \hat{x}$$
$$= \alpha (T^*T + \alpha I)^{-1} (T^*T)^\nu u.$$

Then, using the relation (2.12) in Section 2.3.4, we have

$$\|\hat{x} - x_\alpha\| \le \|u\| \sup_{\lambda \in \sigma(T^*T)} \frac{\alpha \lambda^\nu}{\lambda + \alpha}.$$

Since, for $\lambda > 0$ and $0 < \nu \le 1$,

$$\frac{\lambda^\nu}{\lambda + \alpha} \le \alpha^{\nu-1} \frac{(\lambda/\alpha)^\nu}{1 + \lambda/\alpha} \le \alpha^{\nu-1}$$

it follows that

$$\|\hat{x} - x_\alpha\| \le \|u\| \alpha^\nu.$$

This competes the proof. □

Remark 4.8. The source conditions considered in Theorem 4.12 are particular cases of the assumption $\hat{x} := T^\dagger y \in R((T^*T)^\nu)$ in Theorem 4.15 by taking $\nu = 1/2$ and $\nu = 1$ respectively. Clearly, the condition $T^\dagger y \in R(T^*T)$ is obtained by taking $\nu = 1$. The fact that $T^\dagger y \in R(T^*)$ is obtained by taking $\nu = 1/2$ follows by observing that

$$R(T^*) = R((T^*T)^{1/2})$$

which is a consequence of the *polarization identity* (cf. Kato [31])

$$T = U|T|, \tag{4.32}$$

where $|T| := (T^*T)^{1/2}$. Here, the linear operator $U : X \to X$ is a partial isometry, that is, U satisfies $\|Ux\| = \|x\|$ for all $x \in N(|T|)^\perp$, and U^*U is the orthogonal projection onto $\mathrm{cl}R(|T|)$. Indeed, the polarization identity (4.32) implies $T^* = |T|U^*$ and

$$T^*U = |T|U^*U = (U^*U|T|)^* = |T|^* = |T|$$

so that

$$R(T^*) = R(|T|U^*) \subseteq R(|T|) = R(T^*U) \subseteq R(T^*).$$

Thus, $R(T^*) = R(|T|) = R((T^*T)^{1/2})$.   ◇

4.4.2 *Error estimates with inexact data*

Now, let us consider Tikhonov regularization with an inexact data $\tilde{y}$ in place of the actual data $y \in D(T^\dagger)$. In this case, in place of the regularized solution $x_\alpha := R_\alpha y$, we have $\tilde{x}_\alpha := R_\alpha \tilde{y}$, where R_α is as in (4.23). By Theorem 4.10, we know that $\{R_\alpha\}_{\alpha>0}$ is a regularization family. But, if $R(T)$ is not closed, then $\{\|R_\alpha\| : \alpha > 0\}$ is unbounded (see Corollary 4.6) and by Corollary 4.3, even if $\tilde{y}$ is close to y, the regularized solution $\tilde{x}_\alpha := R_\alpha \tilde{y}$ need not be close to $T^\dagger y$.

As in previous subsection, for $y \in D(T^\dagger)$, we use the notations

$$\hat{x} := T^\dagger y, \quad x_\alpha := R_\alpha y.$$

For $\delta > 0$, let $y^\delta \in Y$ be such that $\|y - y^\delta\| \leq \delta$, and

$$x_\alpha^\delta := R_\alpha y^\delta.$$

Lemma 4.3. *For every $\alpha > 0$ and $\delta > 0$,*

$$\|x_\alpha - x_\alpha^\delta\| \leq \frac{\delta}{2\sqrt{\alpha}}.$$

Proof. Using the estimate for $\|R_\alpha\|$ in Corollary 4.5, we have

$$\|x_\alpha - x_\alpha^\delta\| = \|R_\alpha(y - y^\delta)\|$$
$$\leq \frac{\|y - y^\delta\|}{2\sqrt{\alpha}}$$
$$\leq \frac{\delta}{2\sqrt{\alpha}}.$$

This completes the proof.   □

For the case of $R(T)$ closed, the following theorem is immediate from Theorem 4.7, Corollaries 4.6 and 4.7 and Lemma 4.3.

Theorem 4.16. *Suppose $R(T)$ is closed in Y and $y \in Y$. Then $\hat{x} \in R(T^*T)$ and for every $y^\delta \in Y$*

$$\|\hat{x} - x_\alpha^\delta\| \leq \|u\|\alpha + \|T^\dagger\|\delta,$$

where $u \in X$ is such that $\hat{x} = T^*Tu$.

In particular, if $\alpha \leq c_0\delta$ for some $c_0 > 0$, then

$$\|\hat{x} - x_\alpha^\delta\| \leq c_1\delta,$$

where $c_1 := \|u\|c_0 + \|T^\dagger\|$.

Exercise 4.8. Prove Theorem 4.16.

Exercise 4.9. If $R(T)$ is closed, then show that for every $y, y^\delta \in Y$ with $\|y - y^\delta\| \leq \delta$, $\|\hat{x} - x_\alpha^\delta\| \leq c(\alpha + \delta)$ for some $c > 0$ depending on y.

In the following theorem we specify a parameter choice strategy which ensures convergence.

Theorem 4.17. *The inequality*

$$\|\hat{x} - x_\alpha^\delta\| \leq \|\hat{x} - x_\alpha\| + \frac{\delta}{2\sqrt{\alpha}} \qquad (4.33)$$

holds. In particular, if $\alpha := \alpha(\delta)$ is chosen such that

$$\alpha(\delta) \to 0 \quad and \quad \frac{\delta}{\sqrt{\alpha(\delta)}} \to 0 \quad as \quad \delta \to 0,$$

then

$$\|\hat{x} - x_{\alpha(\delta)}^\delta\| \to 0 \quad as \quad \delta \to 0.$$

Proof. Using the estimate in Lemma 4.3, we have

$$\|\hat{x} - x_\alpha^\delta\| \leq \|\hat{x} - x_\alpha\| + \|x_\alpha - x_\alpha^\delta\|$$
$$\leq \|\hat{x} - x_\alpha\| + \frac{\delta}{2\sqrt{\alpha}}$$

proving (4.33). Also, by Theorem 4.10, $\|\hat{x} - x_\alpha\| \to 0$ as $\alpha \to 0$. Hence, the remaining part of theorem is obvious. $\qquad\square$

Remark 4.9. By Theorem 4.17, a sufficient condition for the Tikhonov regularization to be a regularization method is to have $\alpha = \alpha(\delta)$ such that

$$\alpha(\delta) \to 0 \quad \text{and} \quad \frac{\delta}{\sqrt{\alpha(\delta)}} \to 0 \quad \text{as} \quad \delta \to 0.$$

We may observe that these conditions are satisfied if, for example, $\alpha = c_0 \delta^{2(1-\mu)}$ for some $\mu \in (0,1)$ and for some constant $c_0 > 0$.

Now we give a general error estimate and an order of convergence under a specific smoothness assumption on $\hat{x}$ together with an a priori parameter choice strategy.

Theorem 4.18. *Let $\hat{x} \in R((T^*T)^\nu)$ for some ν with $0 < \nu \leq 1$. Then*

$$\|\hat{x} - x_\alpha^\delta\| \leq c \max\{\alpha^\nu, \delta/\sqrt{\alpha}\},$$

where $c > 0$ is independent of α and δ.

In particular, if $\alpha = c_0 \delta^{2/(2\nu+1)}$ for some constant $c_0 > 0$, then

$$\|\hat{x} - x_\alpha^\delta\| = O(\delta^{2\nu/(2\nu+1)}).$$

Proof. By Theorem 4.15, $\|\hat{x} - x_\alpha\| = O(\alpha^\nu)$, so that by (4.33) we get

$$\|\hat{x} - x_\alpha^\delta\| \leq c \max\{\alpha^\nu, \delta/\sqrt{\alpha}\}$$

for some constant $c > 0$. The particular case is immediate. $\qquad\square$

Remark 4.10. We note that the map

$$\alpha \mapsto \max\{\alpha^\nu, \delta/\sqrt{\alpha}\}, \quad \alpha > 0,$$

attains its minimum at $\alpha = \delta^{2/(2\nu+1)}$, so that the order

$$\|\hat{x} - x_\alpha^\delta\| = O(\delta^{2\nu/(2\nu+1)})$$

given in Theorem 4.18 is the best order possible from the estimate $\max\{\alpha^\nu, \delta/\sqrt{\alpha}\}$. In fact, the above order is sharp. To see this consider the case of a compact operator T with singular system $\{(\sigma_n, u_n, v_n) : n \in \mathbb{N}\}$, and for a fixed $\nu \in (0,1]$ and $k \in \mathbb{N}$, take $y = \sigma_k^{2\nu+1} v_k$. Then, we see that $\hat{x} = \sigma_k^{2\nu} u_k$ and

$$\hat{x} - x_\alpha = \frac{\alpha \sigma_k^{2\nu}}{\sigma_k^2 + \alpha} u_k.$$

Taking $y^\delta := y - \delta v_k$, we have

$$x_\alpha - x_\alpha^\delta = \frac{\delta \sigma_k}{\sigma_k^2 + \alpha} u_k.$$

Now, taking $\delta := \sigma_k^{2\nu+1}$ and $\alpha := \delta^{2/(2\nu+1)}$, it follows that $y^\delta = 0$ so that

$$\|\hat{x} - x_\alpha^\delta\| = \|\hat{x}\| = \sigma_k^{2\nu} = \delta^{2\nu/(2\nu+1)}.$$

This justify our claim. $\qquad\Diamond$

We observe that the best rate for $\|\hat{x} - x_\alpha^\delta\|$ that one obtains from Theorem 4.18 is $O\left(\delta^{2/3}\right)$ which happens if $\nu = 1$. One may ask the following questions:

(i) Can we obtain the order $O(\delta^{2/3})$ by a weaker assumption of α, say for $\alpha := c\delta^{2/3-\varepsilon}$ for some $\varepsilon \in (0, 2/3)$?

(ii) Is it possible to improve the above order $O(\delta^{2/3})$ to

$$\|\hat{x} - x_\alpha^\delta\| = o(\delta^{2/3})$$

possibly by a stronger assumption on $\hat{x}$ and by a different parameter choice strategy?

The following two theorems (cf. Groetsch [25]) show that the answers, in general, are negative.

Theorem 4.19. *If $\hat{x} \neq 0$ and $\alpha := \alpha(\delta, y^\delta)$ is such that $\alpha(\delta, y^\delta) \to 0$ as $d \to 0$, then*

$$\alpha(\delta, y^\delta) = O(\|\hat{x} - x_\alpha^\delta\| + \delta).$$

In particular, if $\|\hat{x} - \tilde{x}_\alpha\| = O(\delta^\mu)$ for some $\mu \in (0, 1]$, then

$$\alpha(\delta, y^\delta) = O(\delta^\mu).$$

Proof. From the definitions of $\hat{x}$ and x_α^δ we have

$$(T^*T + \alpha I)(\hat{x} - x_\alpha^\delta) = \alpha\hat{x} + T^*(y - y^\delta),$$

so that

$$\alpha\|\hat{x}\| \leq (\alpha + \|T^*T\|)\|\hat{x} - x_\alpha^\delta\| + \|T^*\|\delta.$$

Therefore, if $\hat{x} \neq 0$ and if $\alpha := \alpha(\delta, y^\delta)$ is a parameter choice strategy such that $\alpha(\delta, y^\delta) \to 0$ as $\delta \to 0$, then

$$\alpha(\delta, y^\delta) = O(\|\hat{x} - x_\alpha^\delta\| + \delta).$$

The particular case of the result is immediate. $\qquad\square$

Theorem 4.20. *Suppose T is compact with infinite rank, $\hat{x} := T^\dagger y \in R(T^*T)$ and $\alpha := \alpha(\delta, y^\delta)$ is chosen such that*

$$\alpha(\delta, y^\delta) \to 0 \quad as \quad \delta \to 0$$

and

$$\|\hat{x} - x_\alpha^\delta\| = o\left(\delta^{2/3}\right).$$

Then $\hat{x} = 0$.

Proof. Let $\{(\sigma_n, u_n, v_n) : n \in \mathbb{N}\}$ be a singular system of T. Recall that $\sigma_n \to 0$ as $n \to \infty$. Let

$$y_n^\delta = y - \delta v_n, \quad n \in \mathbb{N},$$

so that $\|y - y_n^\delta\| = \delta$. Since $T^* v_n = \sigma_n u_n$ and $T^* T u_n = \sigma_n^2 u_n$, we have

$$R_\alpha(y - y_n^\delta) = \delta(T^*T + \alpha I)^{-1} T^* v_n$$

$$= \frac{\delta \sigma_n}{\sigma_n^2 + \alpha} u_n.$$

Thus,

$$\hat{x} - R_\alpha y_n^\delta = \hat{x} - R_\alpha y + \frac{\delta \sigma_n}{\sigma_n^2 + \alpha} u_n$$

so that

$$\|\hat{x} - R_\alpha y_n^\delta\|^2 = \left\langle (\hat{x} - R_\alpha y) + \frac{\delta \sigma_n}{\sigma_n^2 + \alpha} u_n, \ (\hat{x} - R_\alpha y) + \frac{\delta \sigma_n}{\sigma_n^2 + \alpha} u_n \right\rangle$$

$$\geq \frac{2\delta \sigma_n}{\sigma_n^2 + \alpha} \mathrm{Re}\langle \hat{x} - R_\alpha y, u_n \rangle + \left(\frac{\delta \sigma_n}{\sigma_n^2 + \alpha} \right)^2.$$

Hence,

$$(\delta^{-2/3} \|\hat{x} - R_\alpha y_n^\delta\|)^2 \geq \frac{2\delta^{-1/3} \sigma_n}{\sigma_n^2 + \alpha} \mathrm{Re}\langle \hat{x} - R_\alpha y, u_n \rangle + \left(\frac{\delta^{1/3} \sigma_n^{-1}}{1 + \alpha \sigma_n^{-2}} \right)^2.$$

Let us take $\delta_n := \sigma_n^3$, and let α_n be the corresponding regularization parameter. Then $\delta_n^{-1/3} \sigma_n = 1 = \delta^{1/3} \sigma_n^{-1}$ for all $n \in \mathbb{N}$, so that

$$\left(\delta_n^{-2/3} \|\hat{x} - R_{\alpha_n} y_n^\delta\| \right)^2 \geq \xi_n + \frac{1}{(1 + \alpha \delta_n^{-2/3})^2}, \tag{4.34}$$

where

$$\xi_n := \frac{2\sigma_n^{-2}}{1 + \alpha_n \delta_n^{-2/3}} \mathrm{Re}\langle \hat{x} - R_{\alpha_n} y, u_n \rangle.$$

Since $\hat{x} \in R(T^*T)$, by Theorem 4.12 we have $\|\hat{x} - R_{\alpha_n} y\| \leq \|u\|\alpha$ for some $u \in X$. Hence,

$$|\xi_n| \leq \frac{2\|u\|\alpha_n \sigma_n^{-2}}{1 + \alpha_n \delta_n^{-2/3}} = \frac{2\|u\|\alpha_n \delta_n^{-2/3}}{1 + \alpha_n \delta_n^{-2/3}}. \tag{4.35}$$

Now, suppose $\hat{x} \neq 0$. Then, by Theorem 4.19 we have

$$\alpha(\delta, y^\delta) = O(\|\hat{x} - R_\alpha y^\delta\|),$$

and hence using the hypothesis $\|\hat{x} - R_\alpha y^\delta\| = o(\delta^{2/3})$, we obtain

$$\alpha_n \delta_n^{-2/3} \to 0 \quad \text{as} \quad n \to \infty.$$

Thus, from (4.35) we have $\xi_n \to 0$, and therefore, the right-hand side of (4.34) tends to 1 whereas the left-hand side of (4.34) tends to 0. Thus, we arrive at a contradiction. Therefore, $\hat{x} = 0$. $\qquad \square$

4.4.3 An illustration of the source condition

Now we illustrate the validity of the source condition in Theorem 4.18 in the context of the ill-posed problem associated with the backward heat conduction problem. In Section 4.1.2, we considered the problem of determining the temperature $f := u(\cdot, t_0)$ at a time $t_0 \geq 0$ on a "thin wire" of length ℓ from the knowledge of the temperature $g := u(\cdot, \tau)$ at a time $t = \tau > t_0$, under the assumption that the temperature at both ends of the wire are kept at zero temperature. We have seen that this problem is equivalent to solving a compact operator equation

$$Kf = g,$$

where K is a compact operator with singular values

$$\sigma_n := e^{-\lambda_n^2(\tau - t_0)}, \text{ where } \lambda_n := \frac{cn\pi}{\ell}, \quad n \in \mathbb{N}.$$

In fact, σ_n are the eigenvalues of K with corresponding *eigen functions*

$$\varphi_n(s) := \sqrt{\frac{2}{\ell}} \sin(\lambda_n s), \quad n \in \mathbb{N}.$$

Using the notations in Section 4.1.2, we also know that $f := u(\cdot, t_0)$ must satisfy

$$\langle f, \varphi_n \rangle = e^{-\lambda_n^2 t_0} \langle f_0, \varphi_n \rangle, \quad n \in \mathbb{N},$$

where $f_0 := u(\cdot, 0)$ is the initial temperature. Hence, the condition $f_0 \in L^2[0, \ell]$ implies that

$$\sum_{n=1}^{\infty} e^{\lambda_n^2 t_0} |\langle f, \varphi_n \rangle|^2 = \sum_{n=1}^{\infty} |\langle f_0, \varphi_n \rangle|^2 < \infty.$$

Equivalently,

$$\sum_{n=1}^{\infty} \frac{|\langle f, \varphi_n \rangle|^2}{\sigma_n^{4\nu}} < \infty \quad \text{with} \quad \nu := \frac{t_0}{2(\tau - t_0)}.$$

Thus, in view of the representation (4.31), we see that f satisfies the condition

$$f := K^\dagger g \in R((K^*K)^\nu) \quad \text{with} \quad \nu := \frac{t_0}{2(\tau - t_0)}.$$

From the above, it is clear that, for determining the initial temperature $u(\cdot, 0)$, that is, for $t_0 = 0$, it is necessary to impose additional assumption on $f_0 := u(\cdot, 0)$. Note that if $t_0 = 0$, then the singular values are given by

$$\sigma_n := e^{-\lambda_n^2 \tau}, \quad n \in \mathbb{N}.$$

Hence, the requirement $f_0 := K^\dagger g \in R((K^*K)^\nu)$ for some $\nu > 0$ takes the form

$$\sum_{n=1}^{\infty} e^{4\nu\lambda_n^2 \tau} |\langle f_0, \varphi_n \rangle|^2 < \infty,$$

which is not satisfied automatically by f_0. In order to address this issue, in Section 4.6, we consider a more general type of source condition than the above.

4.4.4 *Discrepancy principles*

A parameter choice strategy in which a priori knowledge on the unknown element $\hat{x}$ required is called an *a priori* parameter choice strategy. For example, the choice of α in Remark 4.9 and Theorem 4.17 (iii) are a priori choices. As against this, in *a posteriori* parameter choice strategies, the parameter α is chosen based on the knowledge of the available data. It has been well recognized that, in applications, an a posteriori parameter choice strategy is preferable to an a priori strategy (cf. [11]). Many a posteriori strategies are available in the literature (cf. [17] and references therein). We consider two of the simplest and oldest a posteriori parameter choice strategies, namely the ones introduced by Morozov and Arcangeli in the year 1966 (cf. [4], [42]) on which the author and his students and collaborators also made some contributions in various contexts (cf. [20, 21, 45, 49, 50, 53, 55–61]).

As in previous subsection, for $y \in D(T^\dagger)$ and $\delta > 0$, let $y^\delta \in Y$ be such that $\|y - y^\delta\| \le \delta$, and

$$\hat{x} := T^\dagger y, \quad x_\alpha := R_\alpha y, \quad x_\alpha^\delta := R_\alpha y^\delta,$$

where $R_\alpha := (T^*T + \alpha I)^{-1} T^*$ for $\alpha > 0$.

(i) *Morozov's and Arcangeli's methods*

In Morozov's method, the parameter α is chosen in such a way that the equation

$$\|T x_\alpha^\delta - y^\delta\| = \delta \tag{4.36}$$

is satisfied, and in Arcangeli's method, the equation to be satisfied is

$$\|T x_\alpha^\delta - y^\delta\| = \frac{\delta}{\sqrt{\alpha}}. \tag{4.37}$$

For proving the existence of unique solutions α in equations (4.36) and (4.37) and also to obtain the corresponding error estimates, we shall make use of a few lemmas.

Lemma 4.4. *Suppose $y \in R(T)$. Then for all $\alpha > 0$,*

$$Tx_\alpha^\delta - y^\delta = -\alpha(TT^* + \alpha I)^{-1}y^\delta, \qquad (4.38)$$

$$\frac{\alpha\|y^\delta\|}{\alpha + \|T\|^2} \leq \|Tx_\alpha^\delta - y^\delta\| \leq \delta + \|\alpha(TT^* + \alpha I)^{-1}y\|, \qquad (4.39)$$

$$\lim_{\alpha \to 0} \|Tx_\alpha^\delta - y^\delta\| \leq \delta \quad and \quad \lim_{\alpha \to \infty} \|Tx_\alpha^\delta - y^\delta\| \geq \|y^\delta\|. \qquad (4.40)$$

Proof. From the definition of x_α^δ we have

$$Tx_\alpha^\delta - y^\delta = T(T^*T + \alpha I)^{-1}T^*y^\delta - y^\delta = -\alpha(TT^* + \alpha I)^{-1}y^\delta.$$

Thus, (4.38) is obtained. Now, since $\|(T^*T + \alpha I)^{-1}\| \leq 1/\alpha$ (cf. Lemma 4.1), we have

$$\|(T^*T + \alpha I)^{-1}y^\delta\| \leq \|(T^*T + \alpha I)^{-1}(y^\delta - y)\| + \|(T^*T + \alpha I)^{-1}y\|$$

$$\leq \frac{\delta}{\alpha} + \|(T^*T + \alpha I)^{-1}y\|.$$

We also have

$$\|y^\delta\| \leq \|T^*T + \alpha I\| \, \|\alpha(T^*T + \alpha I)^{-1}y^\delta\|$$

$$\leq (\alpha + \|T\|^2)\|T\tilde{x} - y^\delta\|.$$

The above two inequalities together with (4.38) give (4.39).

Now, using the estimate $\|(TT^* + \alpha I)^{-1}T\| \leq 1/(2\sqrt{\alpha})$ (see Corollary 4.5) and the fact that $y \in R(T)$, we have

$$\|\alpha(T^*T + \alpha I)^{-1}y\| = \alpha\|(T^*T + \alpha I)^{-1}T\hat{x}\| \leq \frac{\sqrt{\alpha}\|\hat{x}\|}{2}.$$

In particular,

$$\|\alpha(T^*T + \alpha I)^{-1}y\| \to 0 \quad as \quad \alpha \to 0.$$

Hence (4.39) imply (4.40). $\qquad\square$

Lemma 4.5. *Let $A \in \mathcal{B}(X)$ be a positive self-adjoint operator. Then, for $0 \leq \tau \leq 1$,*

$$\|A^\tau x\| \leq \|Ax\|^\tau \|x\|^{1-\tau} \quad \forall x \in X.$$

Proof.   The result is obvious for $\tau = 0$ and for $\tau = 1$. Hence suppose that $0 < \tau < 1$. Now, using the spectral representation of A^τ, we have (see (2.20)),

$$\|A^\tau x\|^2 = \int_a^b \lambda^{2\tau} d\langle E_\lambda x, x\rangle \quad \forall x \in X,$$

where $(a, b) \supseteq \sigma(A)$ and $\{E_\lambda : a < \lambda < b\}$ is the spectral resolution of A. Applying Hölder's inequality (2.3) (see Example 2.5) with $p = 1/\tau$ and $q = 1/(1 - \tau)$, we have

$$\|A^\tau x\|^2 \leq \left(\int_a^b \lambda^2 d\langle E_\lambda x, x\rangle\right)^\tau \left(\int_a^b d\langle E_\lambda x, x\rangle\right)^{1-\tau}$$
$$= \|Ax\|^{2\tau}\|x\|^{2(1-\tau)} \quad \forall x \in X,$$

so that the result follows.   □

The inequality in Lemma 4.5 is called an *interpolation inequality*.

For the next lemma we use the fact that if $\nu \in (0, 1]$, then $R((T^*T)^\nu) \subseteq R((T^*T)^\mu)$, where $\mu := \min\{\nu, 1/2\}$.

Exercise 4.10. Justify the above statement.

Lemma 4.6. *Suppose $\hat{x} \in R((T^*T)^\nu)$ for some $\nu \in (0, 1]$. Then for every $\alpha > 0$,*

$$\|\alpha(TT^* + \alpha I)^{-1}T\hat{x}\| \leq \|u\|\alpha^{\mu+1/2},$$

where $\mu := \min\{\frac{1}{2}, \nu\}$ and $u \in X$ is such that $\hat{x} = (T^*T)^\mu u$.

Proof.   Using the spectral representation (2.19), we have

$$(TT^* + \alpha I)^{-1}T\hat{x} = (TT^* + \alpha I)^{-1}T(T^*T)^\mu u$$
$$= T(T^*T + \alpha I)^{-1}(T^*T)^\mu u,$$

and

$$\|(TT^* + \alpha I)^{-1}T\hat{x}\|^2 = \langle(T^*T + \alpha I)^{-2}(T^*T)^{2\mu+1}u, u\rangle$$
$$= \|u\|^2\|(T^*T + \alpha I)^{-2}(T^*T)^{2\mu+1}\|.$$

Hence, using the inequality (2.12) (cf. Section 2.3.4), we obtain

$$\|\alpha(TT^* + \alpha I)^{-1}T\hat{x}\| \leq \|u\|\alpha^{\mu+1/2}.$$

This completes the proof.   □

Lemma 4.7. *Let* $0 \neq y \in D(T^{\dagger})$. *Suppose that* $\alpha := \alpha(\delta, y^{\delta})$ *is chosen such that* $\alpha^{r}\|Tx_{\alpha}^{\delta} - y^{\delta}\| \to 0$ *as* $\delta \to 0$ *for some* $r \in \mathbb{R}$. *Then* $\alpha(\delta, y^{\delta}) \to 0$ *as* $\delta \to 0$.

Proof. Suppose the conclusion in the lemma does not hold. Then, there exists a sequence (δ_{n}) of positive real numbers which converges to 0 and a sequence (y_{n}) in Y such that $\|y - y_{n}\| \leq \delta_{n}$ for all $n \in \mathbb{N}$, but $\alpha_{n} := \alpha(\delta_{n}, y_{n}) \to c$ for some $c > 0$. Now, from the relation (4.38) in Lemma 4.4, for any $r > 0$ we obtain

$$\lim_{n \to \infty} \alpha_{n}^{r}\|Tx_{\alpha_{n}}^{\delta_{n}} - y^{\delta_{n}}\| = c^{r}\|(TT^{*} + cI)^{-1}y\| \neq 0.$$

This is a contradiction to the hypothesis of the lemma. $\square$

Theorem 4.21. *Let* $0 \neq y \in D(T^{\dagger})$. *Then equations (4.36) and (4.37) have unique solutions, say* $\alpha^{(M)}(\delta, y^{\delta})$ *and* $\alpha^{(A)}(\delta, y^{\delta})$, *respectively. Moreover,*

$$\alpha^{(M)}(\delta, y^{\delta}) \to 0 \quad and \quad \alpha^{(A)}(\delta, y^{\delta}) \to 0 \quad as \quad \delta \to 0,$$

and

$$\alpha^{(M)}(\delta, y^{\delta}) = O(\delta) \quad and \quad \alpha^{(A)}(\delta, y^{\delta}) = O(\delta^{2/3}).$$

Proof. It is seen that the map

$$\alpha \mapsto \|Tx_{\alpha}^{\delta} - y^{\delta}\|^{2} = \alpha^{2}\langle(TT^{*} + \alpha I)^{-2}y^{\delta}, y^{\delta}\rangle, \quad \alpha > 0,$$

is continuous and strictly increasing, so that the existence of a unique $\alpha = \alpha(\delta, y^{\delta})$ satisfying (4.36) (resp. (4.37)) follows from (4.40) by using intermediate value theorem. Hence, by Lemma 4.7, taking $r = 0$ and $r = 1/2$, respectively, we have

$$\alpha^{(M)}(\delta, y^{\delta}) \to 0 \quad and \quad \alpha^{(A)}(\delta, y^{\delta}) \to 0 \quad as \quad \delta \to 0.$$

Now, writing the relation (4.39) in Lemma 4.4 as

$$\alpha \leq \frac{(\alpha + \|T\|^{2})\|Tx_{\alpha}^{\delta} - y^{\delta}\|}{\|y^{\delta}\|}$$

and using the facts that $\alpha^{(M)}(\delta, y^{\delta}) \to 0$ and $\alpha^{(A)}(\delta, y^{\delta}) \to 0$ as $\delta \to 0$, we have $\alpha^{(M)}(\delta, y^{\delta}) = O(\delta)$ and $\alpha^{(A)}(\delta, y^{\delta}) = O(\delta^{2/3})$. $\square$

Theorem 4.22. *Let* $0 \neq y \in R(T)$ *and* $\hat{x} \in R(T^{*}T)^{\nu}$ *for some* $\nu \in (0, 1]$. *Then the following hold.*

(i) *If α is chosen according to the Morozov's discrepancy principle (4.36), then*

$$\|\hat{x} - x_\alpha^\delta\| \le \|u\|^{\frac{1}{2\mu+1}} \delta^{\frac{2\mu}{2\mu+1}}, \tag{4.41}$$

where $\mu = \min\{\nu, 1/2\}$ and $u \in X$ is such that $\hat{x} = (T^*T)^\mu u$.

(ii) *If α is chosen according to the Arcangeli's discrepancy principle (4.37), then*

$$\|\hat{x} - x_\alpha^\delta\| = O(\delta^{\frac{2\nu}{3}}). \tag{4.42}$$

Proof.   (i) Suppose α is chosen according to (4.36). The we have

$$\begin{aligned}
\delta^2 + \alpha\|x_\alpha^\delta\|^2 &= \|Tx_\alpha^\delta - y^\delta\|^2 + \alpha\|x_\alpha^\delta\|^2 \\
&\le \|T\hat{x} - y^\delta\|^2 + \alpha\|\hat{x}\|^2 \\
&\le \delta^2 + \alpha\|\hat{x}\|^2,
\end{aligned}$$

so that $\|x_\alpha^\delta\| \le \|\hat{x}\|$. Therefore,

$$\begin{aligned}
\|\hat{x} - x_\alpha^\delta\|^2 &= \langle \hat{x} - x_\alpha^\delta, \hat{x} - x_\alpha^\delta \rangle \\
&= \|\hat{x}\|^2 - 2Re\langle \hat{x}, x_\alpha^\delta \rangle + \|x_\alpha^\delta\|^2 \\
&\le 2\|\hat{x}\|^2 - 2Re\langle \hat{x}, x_\alpha^\delta \rangle \\
&= \langle \hat{x}, \hat{x} - x_\alpha^\delta \rangle + \langle \hat{x} - x_\alpha^\delta, \hat{x} \rangle \\
&\le 2|\langle \hat{x} - x_\alpha^\delta, \hat{x} \rangle|.
\end{aligned}$$

Now, assume that $\hat{x} \in R((T^*T)^\nu)$ for some ν, and let $u \in X$ be such that $\hat{x} = (T^*T)^\mu u$, where $\mu = \min\{1/2, \nu\}$. Then we have

$$\begin{aligned}
\|\hat{x} - x_\alpha^\delta\|^2 &\le 2\langle (T^*T)^\mu(\hat{x} - x_\alpha^\delta), u \rangle \\
&\le 2\|(T^*T)^\mu(\hat{x} - x_\alpha^\delta)\|\|u\|.
\end{aligned}$$

Since $\|(T^*T)^{1/2}x\| = \|Tx\|$ for all $x \in X$, it follows from Lemma 4.5 by taking $A = (T^*T)^{1/2}$ and $\tau = 2\mu$ that

$$\|(T^*T)^\mu x\| \le \|Tx\|^{2\mu}\|x\|^{1-2\mu} \tag{4.43}$$

for any $x \in X$. Therefore, we obtain

$$\|\hat{x} - x_\alpha^\delta\|^2 \le 2\|u\|\|T(\hat{x} - x_\alpha^\delta)\|^{2\mu}\|\hat{x} - x_\alpha^\delta\|^{1-2\mu},$$

so that

$$\|\hat{x} - x_\alpha^\delta\|^{2\mu+1} \le 2\|u\|\|T(\hat{x} - x_\alpha^\delta)\|^{2\mu}.$$

But,

$$\|T(\hat{x} - x_\alpha^\delta)\| = \|y - Tx_\alpha^\delta\| \le \|y - y^\delta\| + \|y^\delta - Tx_\alpha^\delta\| \le 2\delta.$$

Therefore,

$$\|\hat{x} - x_\alpha^\delta\| \le (2\|u\|)^{1/(2\mu+1)}(2\delta)^{2\mu/(2\mu+1)}$$
$$= \|u\|^{1/(2\mu+1)}\delta^{2\mu/(2\mu+1)}.$$

Thus, proof of (i) is over.

(ii) Now, let α satisfy (4.37). Then by Lemma 4.6 and the relation (4.39), we have

$$\frac{\delta}{\sqrt{\alpha}} = \|Tx_\alpha^\delta - y^\delta\|$$
$$\le \delta + \|\alpha(TT^* + \alpha I)^{-1}T\hat{x}\|$$
$$\le \delta + \|u\|\alpha^\omega,$$

where $\omega := \min\{1, \nu + 1/2\}$. Hence, using the fact that $\alpha = O(\delta^{2/3})$, it follows that $\delta/\sqrt{\alpha} = O(\delta^{2\omega/3})$. Therefore, Theorem 4.17 (ii) gives

$$\|\hat{x} - x_\alpha^\delta\| = O(\max\{\alpha^\nu, \delta/\sqrt{\alpha}\})$$
$$= O(\delta^{2\nu/3}).$$

This completes the proof of the theorem. $\qquad\square$

Remark 4.11. By Theorem 4.22, we see that if $\hat{x} \in R((T^*T)^\nu)$ with $0 < \nu \le 1$, then the optimal rate

$$\|\hat{x} - x_\alpha^\delta\| = O\left(\delta^{\frac{2\nu}{2\nu+1}}\right)$$

is obtained under Morozov's discrepancy principle if $0 < \nu \le 1/2$, and under Arcangeli's discrepancy principle if $\nu = 1$. In fact, it is known that, under Morozov's discrepancy principle, it is not possible to get a rate better than $O\left(\delta^{1/2}\right)$ (cf. Groetsch [24]).

We may also observe that

$$\delta^{2\nu/3} > \delta^{1/2} \quad \text{if and only if} \quad 3/4 < \nu \le 1,$$

so that the rate obtained under Arcangeli's method is better than the rate under Morozov's method for $3/4 < \nu \le 1$, whereas for $0 < \nu < 3/4$, the rate under Arcangeli's method is not as good as that under Morozov's method.

We may also recall from Section 4.4.2 that under Tikhonov regularization we cannot obtain a rate better than $O(\delta^{2/3})$; whatever be the mode of parameter choice strategy. $\qquad\diamond$

Remark 4.12. Looking at the proof of Theorem 4.22, it can be seen that results in (i) and (ii) in Theorem 4.22 hold with α satisfying

$$\delta \leq \|Tx_\alpha^\delta - y^\delta\| \leq c_1 \delta$$

and

$$\frac{\delta}{\sqrt{\alpha}} \leq \|Tx_\alpha^\delta - y^\delta\| \leq c_2 \frac{\delta}{\sqrt{\alpha}}$$

for some $c_1, c_2 \geq 1$ instead of (4.36) and (4.37) respectively. $\Diamond$

Exercise 4.11. Verify the assertions in Remark 4.12.

A natural query, at this point, would be whether we can modify the proof Theorem 4.22 so as to obtain a rate better than $O(\delta^{1/2})$ under Morozov's method for $1/2 \leq \nu \leq 1$. The following result due to Groetsch ([24]) shows that it is not possible.

Theorem 4.23. *If $T : X \to Y$ is a compact operator of infinite rank, then the best rate possible under Morozov's discrepancy principle (4.36) is $O(\delta^{1/2})$.*

Proof.   Let $\{(\sigma_n, u_n, v_n) : n \in \mathbb{N}\}$ be a singular system for T. Let us assume that under Morozov's discrepancy principle (4.36), it is possible to get the rate

$$\|\hat{x} - x_a^\delta\| = o(\delta^{1/2})$$

whenever $\hat{x} \in R(T^*T)$. We shall arrive at a contradiction.

Let (δ_n) be a sequence of positive reals such that $\delta_n \to 0$ as $n \to \infty$, and let

$$y = v_1, \qquad y_n = y - \delta_n v_n \quad \forall n \in \mathbb{N}.$$

In this case we have $\hat{x} := T^\dagger y = u_1/\sigma_1$. Clearly $\|y - y_n\| = \delta_n$ for every $n \in \mathbb{N}$. For each fixed $n \in \mathbb{N}$, let $\alpha_n = \alpha(\delta_n, y_n)$ be the regularization parameter obtained under the Morozov's discrepancy principle (4.36), and let

$$x_n = x_{\alpha_n}, \qquad x_n^\delta = x_{\alpha_n}^\delta.$$

Since $(T^*T + \alpha_n I)^{-1} u_n = u_n/(\sigma_n^2 + \alpha_n)$ for all $n \in \mathbb{N}$, we have

$$\hat{x} - x_n^\delta = (\hat{x} - x_n) + (x_n - x_n^\delta)$$

$$= \alpha_n (T^*T + \alpha_n I)^{-1} \hat{x} + (T^*T + \alpha_n I)^{-1} T^* (y - y_n)$$

$$= \frac{\alpha_n}{\sigma_1} (T^*T + \alpha_n I)^{-1} u_1 + \delta_n (T^*T + \alpha_n I)^{-1} T^* v_n$$

$$= \frac{\alpha_n}{\sigma_1 (\sigma_n^2 + \alpha_n)} u_1 + \frac{\delta_n \sigma_n}{\sigma_n^2 + \alpha_n} u_n.$$

Hence for $n > 1$,

$$\|\hat{x} - x_n^\delta\|^2 = \frac{\alpha_n^2}{\sigma_1^2(\sigma_n^2 + \alpha_n)^2} + \frac{\delta_n^2 \sigma_n^2}{(\sigma_n^2 + \alpha_n)^2}$$

$$\geq \left(\frac{\delta_n \sigma_n}{\sigma_n^2 + \alpha_n}\right)^2.$$

Now, taking $\delta_n = \sigma_n^2$ we have

$$\|\hat{x} - x_n^\delta\| \geq \frac{\delta_n \sigma_n}{\sigma_n^2 + \alpha_n} = \frac{\delta_n^{3/2}}{\delta_n + \alpha_n}.$$

Thus, the assumption $\|\hat{x} - x_\alpha^\delta\| = o(\delta^{1/2})$ implies

$$\frac{\delta_n}{\delta_n + \alpha_n} \leq \frac{\|\hat{x} - x_n^\delta\|}{\sqrt{\delta_n}} \to 0 \quad \text{as} \quad n \to \infty.$$

Consequently, we get

$$\frac{\alpha_n}{\delta_n} \to \infty \quad \text{as} \quad n \to \infty.$$

This is a contradiction to the fact that $\alpha_n = O(\delta_n)$ obtained in Theorem 4.21. $\qquad\Box$

Our next attempt is to look for an a posteriori parameter choice strategy which gives a better rate than $O(\delta^{2\nu/3})$ for $0 < \nu < 1$. In this regard, we consider a modified form of Arcangeli's method defined using a pair (p, q) of parameters.

(ii) Generalized Arcangeli's method

For $p > 0$ and $q > 0$, consider the equation

$$\|Tx_\alpha^\delta - y^\delta\| = \frac{\delta^p}{\alpha^q} \tag{4.44}$$

for choosing the regularization parameter α.

Let $\delta_0 > 0$ be such that $\|y\| \geq 2\delta_0$ and for $\delta \in (0, \delta_0]$, let $y^\delta \in Y$ be such that

$$\|y - y^\delta\| \leq \delta.$$

Theorem 4.24. *For each $\delta \in (0, \delta_0]$, the equation (4.44) has a unique solution $\alpha = \alpha(\delta, y^\delta)$. Moreover*

$$\alpha(\delta, y^\delta) \to 0 \quad as \quad \delta \to 0$$

and

$$\alpha(\delta, y^\delta) \leq c_1 \delta^{\frac{p}{q+1}} \tag{4.45}$$

for some $c_1 > 0$. If in addition $y \in R(T)$, $\hat{x} \in R(T^*T)^\nu$ for some $\nu \in (0,1]$ and

$$\frac{p}{q+1} \leq \min\left\{1, \frac{4q}{2q+1}\right\}, \tag{4.46}$$

then

$$\frac{\delta}{\sqrt{\alpha}} = O\left(\delta^\mu\right),$$

where

$$\mu = 1 - \frac{p}{2(q+1)}\left(1 + \frac{1-\omega}{q}\right) \quad \text{with} \quad \omega = \min\left\{1, \nu + \frac{1}{2}\right\}.$$

Proof. The existence of a unique $\alpha = \alpha(\delta, y^\delta)$ such that (4.44) is satisfied follows as the case of Morozov's and Arcangeli's discrepancy principles (cf. Theorem 4.21), that is, by using the fact that the map

$$\alpha \mapsto \|Tx_\alpha^\delta - y^\delta\|^2 = \alpha^2 \langle (TT^* + \alpha I)^{-2} y^\delta, y^\delta \rangle, \quad \alpha > 0,$$

is continuous and strictly increasing and then appealing to intermediate value theorem. By Lemma 4.7, we also obtain the convergence $\alpha(\delta, y^\delta) \to 0$ as $\delta \to 0$. Now, recall from Lemma 4.4 that

$$\frac{\alpha \|y^\delta\|}{\alpha + \|T\|^2} \leq \|Tx_\alpha^\delta - y^\delta\| \leq \delta + \|\alpha(TT^* + \alpha I)^{-1} y\|. \tag{4.47}$$

The first inequality in (4.47) and the convergence $\alpha(\delta, y^\delta) \to 0$ as $\delta \to 0$ imply that $\alpha(\delta, y^\delta) \leq c_1 \delta^{\frac{p}{q+1}}$ for some $c_1 > 0$. Also, the second inequality in (4.47) together with Lemma 4.6 implies that

$$\frac{\delta^p}{\alpha^q} = \|Tx_\alpha^\delta - y^\delta\| \leq \delta + \|u\|\alpha^\omega,$$

where $u \in X$ is such that $\hat{x} = (T^*T)^\nu u$. Therefore, by the estimate in (4.45),

$$\frac{\delta^p}{\alpha^q} \leq \delta + c_1^\omega \|u\| \delta^{\frac{p\omega}{q+1}}.$$

From this we have

$$\frac{\delta}{\sqrt{\alpha}} = \delta^{1 - \frac{p}{2q}} \left(\frac{\delta^p}{\alpha^q}\right)^{\frac{1}{2q}}$$

$$\leq \left(\delta^{2q-p+1} + c_1^\omega \|u\| \delta^{2q-p+\frac{p\omega}{q+1}}\right)^{1/2q}.$$

Using the condition (4.46) on (p,q), we have

$$0 \leq 2q - p + \frac{p}{2(q+1)} \leq 2q - p + \frac{p\omega}{q+1} \leq 2q - p + 1.$$

Hence, it follows that

$$\frac{\delta}{\sqrt{\alpha}} \leq c_2 \delta^\mu$$

for some $c_2 > 0$, where

$$\mu = \frac{1}{2q}\left(2q - p + \frac{p\omega}{q+1}\right) = 1 - \frac{p}{2(q+1)}\left(1 + \frac{1-\omega}{q}\right).$$

This completes the proof. □

Theorem 4.25. *Suppose* $\dfrac{p}{q+1} \leq \min\left\{1, \dfrac{4q}{2q+1}\right\}$ *and* α *is chosen according to (4.44). If* $\hat{x} \in R((T^*T)^\nu)$ *for some* $\nu \in (0,1]$, *then*

$$\|\hat{x} - x_\alpha^\delta\| \leq c_0 \delta^t$$

for some $c_0 > 0$, *where*

$$t = \min\left\{\frac{p\nu}{q+1}, \; 1 - \frac{p}{2(q+1)}\left(1 + \frac{1-\omega}{q}\right)\right\}.$$

In particular, if $p \leq 3/2$, $q \geq 1/2$ *and*

$$\frac{p}{q+1} \leq \frac{2}{2\nu + 1 + \frac{1-\omega}{q}},$$

then

$$\|\hat{x} - x_\alpha^\delta\| = O(\delta^{\frac{p\nu}{q+1}}).$$

Proof. Under the assumption on $\hat{x}$, from Theorem 4.17 we have

$$\|\hat{x} - x_\alpha^\delta\| \leq c \max\left\{\alpha^\nu, \frac{\delta}{\sqrt{\alpha}}\right\},$$

so that the result follows from Theorem 4.24. We obtain the particular case by observing that

$$\frac{4q}{2q+1} \leq 1 \iff q \geq \frac{1}{2};$$

$$p \leq \frac{3}{2}, q \geq \frac{1}{2} \implies \frac{p}{q+1} \leq 1$$

and

$$\frac{p\nu}{q+1} \leq 1 - \frac{p}{2(q+1)}\left(1 + \frac{1-\omega}{q}\right) \iff \frac{p}{q+1} \leq \frac{2}{2\nu + 1 + \frac{1-\omega}{q}}.$$

This completes the proof. □

Remark 4.13. Since $p/(q+1)$ is assumed to be less than 1, the order $O(\delta^t)$ that results from Theorem 4.25 is not as good as $O(\delta^\nu)$, and hence the rate is not as good as $O(\delta^{2\nu/(2\nu+1)})$ for $0 < \nu < 1/2$. $\Diamond$

The following corollary is immediate from Theorem 4.25.

Corollary 4.8. *Suppose $\hat{x} \in R(T^*T)^\nu$ with $0 < \nu \leq 1$ and α is chosen according to (4.44) where $p/(q+1) = 2/3$ with $p \leq 3/2$ and $q \geq 1/2$. Then*

$$\|\hat{x} - x_\alpha^\delta\| \leq c_0 \delta^{\frac{2\nu}{3}}$$

for some $c_0 > 0$.

Exercise 4.12. Supply details for the proof of Corollary 4.8.

Remark 4.14. Corollary 4.8 shows that the estimate

$$\|\hat{x} - x_\alpha^\delta\| = O(\delta^{\frac{2\nu}{3}})$$

is not only valid for the Arcangeli's method, that is for $p = 1$, $q = 1/2$, but for a class of discrepancy principles (4.44) with p and q satisfying

$$\frac{p}{q+1} = \frac{2}{3}, \quad p \leq \frac{3}{2}, \quad q \geq \frac{1}{2}.$$

Moreover, from Theorem 4.25, we can infer that if $\hat{x} \in R(T^*T)^\nu$, $\nu \geq 1/2$, and if we know some upper estimate ν_0 of ν, i.e.,

$$\frac{1}{2} \leq \nu_0 \leq 1 \quad \text{and} \quad 1/2 \leq \nu \leq \nu_0,$$

then by taking

$$\frac{p}{q+1} = \frac{2}{2\nu_0 + 1}$$

with $p \leq 3/2$ and $q \geq 1/2$, for example, taking $p = 1$ and $q = \nu_0 + 1/2$ or $p = 2$ and $q = 2\nu_0 + 1$, we obtain

$$\|\hat{x} - x_\alpha^\delta\| = O(\delta^{\frac{2\nu}{2\nu_0+1}}).$$

It is still an open question whether the Arcangeli's method can lead to the rate $O(\delta^{2\nu/(2\nu+1)})$ when $\hat{x} \in R((T^*T)^\nu)$ for any $\nu \in (0,1)$. $\Diamond$

Remark 4.15. The discrepancy principle (4.44) was first considered by Schock [70], and obtained the rate

$$\|\hat{x} - x_\alpha^\delta\| = O(\delta^s), \quad s := \frac{2\nu}{2\nu + 1 + \frac{1}{2q}}$$

under the assumption

$$\frac{p}{q+1} = \frac{2}{2\nu + 1 + \frac{1}{2q}}.$$

This result was a generalized form of the earlier known result for Arcangeli's method (cf. Groetsch and Schock [28]), namely,

$$\|\hat{x} - x_\alpha^\delta\| = O(\delta^{1/3}).$$

Later, Nair [45] improved the result of Schock and proved that if

$$\frac{p}{q+1} = \frac{2}{2\nu + 1 + (1 - \nu)/2q}$$

then

$$\|\hat{x} - x_\alpha^\delta\| = O(\delta^r), \quad r := \frac{2\nu}{2\nu + 1 + (1 - \nu)/2q}.$$

This result not only improves the result of Schock, but also gives the best rate

$$\|\hat{x} - x_\alpha^\delta\| = O(\delta^{2/3})$$

for a class of discrepancy principles (4.44) with $p/(q+1) = 2/3$. Theorem 4.25 is an improved form of the above result in [45], and it is proved by George and Nair [20]. $\Diamond$

4.4.5 *Remarks on general regularization*

Tikhonov regularization is a special case of a general regularization given by

$$R_\alpha := g_\alpha(T^*T)T^*, \quad \alpha > 0,$$

where g_α is a piecewise continuous non-negative real valued function defined on $[0, a]$ where $a \geq \|T\|^2$ (cf. [17]). Here, the operator $g_\alpha(T^*T)$ is defined via spectral theorem, that is,

$$g_\alpha(T^*T) = \int_0^a g_\alpha(\lambda) dE_\lambda,$$

where $\{E_\lambda : 0 \leq \lambda \leq a\}$ is the resolution of identity for the self adjoint operator T^*T. Note that the choice $g_\alpha(\lambda) = 1/(\lambda + \alpha)$ corresponds to the Tikhonov regularization. Of course, for the above $\{R_\alpha\}_{\alpha>0}$ to be a regularization family, we have to impose certain conditions on g_α, $\alpha > 0$. In this regard, we have the following.

Theorem 4.26. *Suppose there exists $c > 0$ such that*

$$|1 - \lambda g_\alpha(\lambda)| \le c \quad \forall \lambda \in [0, a], \, \alpha > 0,$$

and

$$\xi_\alpha := \sup \sqrt{\lambda}|1 - \lambda g_\alpha(\lambda)| \to 0 \quad as \quad \alpha \to 0.$$

Then $\{R_\alpha\}_{\alpha>0}$ to be a regularization family.

Proof.   For $y \in D(T^\dagger)$, let $\hat{x} := T^\dagger y$. Then we have

$$\hat{x} - R_\alpha y = \hat{x} - g_\alpha(T^*T)T^*y$$
$$= [I - g_\alpha(T^*T)T^*T]\hat{x}.$$

Let $A_\alpha := I - g_\alpha(T^*T)T^*T$. Then we have

$$\|A_\alpha\| \le \sup_{0<\lambda\le a} |1 - \lambda g_\alpha(\lambda)| \le c \quad \forall \alpha > 0,$$

and for $x \in R(T^*)$, if $u \in Y$ is such that $x = T^*u$, then

$$\|A_\alpha x\| = \|A_\alpha T^*u\|$$
$$\le \|u\| \sup_{0<\lambda\le a} \sqrt{\lambda}|1 - \lambda g_\alpha(\lambda)|$$
$$= \|u\|\xi_\alpha$$
$$\to 0 \quad as \quad \alpha \to 0.$$

Now, since $\hat{x} \in N(T)^\perp$ and $R(T^*)$ is dense in $N(T)^\perp$, it follows from Theorem 2.8 that $\|\hat{x} - R_\alpha y\| = \|A_\alpha \hat{x}\| \to 0$ as $\alpha \to 0$. $\square$

Some of the special cases of the general regularization method other than the Tikhonov regularization are the iterated Tikhonov regularization, asymptotic regularization or Schwalter's method, truncated spectral method.

(i) In **iterated Tikhonov regularization**, the regularized solution $x_{\alpha,m}$ for $m \in \mathbb{N}$, is defined iteratively as the solution of

$$(T^*T + \alpha I)x_{\alpha,k} = \alpha x_{k-1} + T^*y, \quad k = 1, \dots, m.$$

In this case, it can be seen that $x_{\alpha,m} = g_\alpha(T^*T)T^*y$ with

$$g_\alpha(\lambda) := \frac{1}{\lambda}\Big[1 - \Big(\frac{\alpha}{\lambda + \alpha}\Big)^m\Big].$$

(ii) In **asymptotic regularization**, the regularized solution x_α is defined as $x_\alpha := u_\alpha(1/\alpha)$, where $u(\cdot)$ is the solution of the initial value problem,

$$u'(t) + T^*Tu(t) = T^*y, \quad u(0) = 0, \quad t \in (0, \infty).$$

In this case, it can be seen that

$$g_\alpha(\lambda) := \frac{1}{\lambda}(1 - e^{-\lambda/\alpha}).$$

(iii) In **truncated spectral method**, g_α is defined by

$$g_\alpha(\lambda) = \begin{cases} 1/\lambda, & \lambda \geq \alpha, \\ 0, & \lambda < \alpha. \end{cases}$$

Thus, the regularized solution x_α is given by

$$x_\alpha = \left(\int_\alpha^a \frac{1}{\lambda} dE_\alpha \right) T^* y.$$

If T is a compact operator with singular system $\{(\sigma_n, u_n, v_n) : n \in \mathbb{N}\}$, then x_α takes the form

$$x_\alpha = \sum_{n \in \Lambda_\alpha} \frac{\langle y, v_n \rangle}{\sigma_n} u_n,$$

where $\Lambda_\alpha := \{n \in \mathbb{N} : \sigma_n^2 > \alpha\}$.

It can be seen that for Tikhonov regularization and in all the above three methods, the conditions in Theorem 4.26 are satisfied.

4.5 Best Possible Worst Case Error

We have seen that under the assumption of $\hat{x} \in R((T^*T)^\nu)$, the best rate possible under Tikhonov regularization is $O(\delta^\mu)$ where $\mu = \min\{\frac{2\nu}{2\nu+1}, 1\}$ (see Remark 4.10 in Section 4.4.2). Now, we would like to know the best rate possible under any regularization method whenever $\hat{x}$ belongs to a subset of $R((T^*T)^\nu)$, $\nu > 0$. In fact, we shall address the above question not only for the above situation, but also when $\hat{x}$ belongs to a certain subset of $R([\varphi(T^*T)]^{1/2})$, where $\varphi : [0, a] \to [0, \infty)$ for $a \geq \|T\|^2$ is a continuous function satisfying certain additional conditions.

First let us define the concept of *order optimal algorithms* in terms of a quantity which can be considered as a measure of order optimality with respect to a *source set* $M \subseteq X$, irrespective of the regularization that we use.

A map $R : Y \to X$ is called a **reconstruction algorithm** for solving equation (4.1), and corresponding to the inexact data $y^\delta \in Y$, Ry^δ can be thought of as an approximation of $\hat{x}$. For example, $R := R_\alpha$ can be an operator associated with a regularization method with certain parameter

α. Given an algorithm R, a *source set* $M \subseteq X$ and an error $\delta > 0$, the quantity

$$E_R(M, \delta) := \sup\{\|x - Ry^\delta\| : x \in M, \|Tx - y^\delta\| \le \delta\}$$

is called the **worst case error** associated with the triple (R, M, δ). Let $\mathcal{R}$ be the family of all reconstruction algorithms for equation (4.1), and

$$E(M, \delta) := \inf_{R \in \mathcal{R}} E_R(M, \delta).$$

Then the effort would be to have a regularization method R_0 such that for any $y \in D(T^\dagger)$ and $y^\delta \in Y$ with $\|y - y^\delta\| \le \delta$,

$$\|\hat{x} - R_0 y^\delta\| \le c_0 E(M, \delta), \tag{4.48}$$

for some $c_0 \ge 1$. Such an algorithm, if exists, is called an **order optimal algorithm** for equation (4.1), and the quantity $E(M, \delta)$ is called the **best possible worst case error** for (4.1).

Let us define another quantity associated with the source set M, namely,

$$\omega(M, \delta) := \sup\{\|x\| : x \in M, \|Tx\| \le \delta\}. \tag{4.49}$$

It is obvious from the definition of $E_R(M, \delta)$ that if R is a linear operator, then

$$E_R(M, \delta) \ge \omega(M, \delta).$$

Exercise 4.13. Prove the above statement.

Thus, if R_0 is a linear regularization method for (4.1) for which

$$\|\hat{x} - R_0 y^\delta\| \le c_0 \omega(M, \delta), \tag{4.50}$$

holds for some $c_0 > 0$, then

$$\|\hat{x} - R_0 y^\delta\| \le c_0 \inf_{R \in \mathcal{L}(Y,X)} E_R(M, \delta),$$

where $\mathcal{L}(Y, X)$ denotes the space of all linear operators from Y to X. The following proposition due to Miccelli and Rivlin [41] shows that we, in fact, have (4.48) whenever (4.50) holds, provided the source set satisfies $\{-x : x \in M\} \subseteq M$.

Proposition 4.4. *Suppose $M \subset X$ satisfies $\{-x : x \in M\} \subseteq M$. Then*

$$\omega(M, \delta) \le E(M, \delta).$$

Proof. Suppose $x \in M$ such that $\|Tx\| \leq \delta$ and $R : Y \to X$ be any arbitrary map. Then,

$$\|x - R(0)\| \leq E_R(M, \delta).$$

Since $-x \in M$ and $\|T(-x)\| = \|Tx\| \leq \delta$, we also have

$$\|x + R(0)\| = \| - x - R(0)\| \leq E_R(M, \delta).$$

Thus,

$$2\|x\| = \|2x\| = \|(x - R(0)) + (x + R(0))\| \leq 2\, E_R(M, \delta).$$

Since this is true for all $R : Y \to X$, we have $\|x\| \leq E(M, \delta)$ for all $x \in M$ with $\|Tx\| \leq \delta$. Hence, $\omega(M, \delta) \leq E(M, \delta)$. $\qquad\square$

Remark 4.16. It has also been shown in [41] that if M is *balanced* and *convex*, then

$$E(M, \delta) \leq 2\,\omega(M, \delta).$$

We may recall that a subset E of a vector space is said to be a **balanced set** if $\lambda x \in E$ whenever $x \in E$ and $\lambda \in \mathbb{K}$ such that $|\lambda| \leq 1$, and E is said to be a **convex set** if $\lambda x + (1 - \lambda)y \in E$ whenever $x, y \in E$ and $\lambda \in [0, 1]$.

Note that, if T is injective, then the quantity $\omega(M, \delta)$ can be thought of as the **modulus of continuity** of $T^{-1} : T(M) \to X$. $\qquad\lozenge$

If we are looking for a regularization method R_0 satisfying (4.49), then it is apparent that the requirement on the source set M should be such that

$$\omega(M, \delta) \to 0 \quad \text{as} \quad \delta \to 0. \tag{4.51}$$

In this context, it is to be observed that if T has continuous inverse, then for any M,

$$\omega(M, \delta) \leq \|T^{-1}\|\delta$$

so that the requirement (4.51) is met.

Exercise 4.14. Justify the above statement.

The following proposition shows that if T is not injective and if M contains a non-zero element from the null space of T, then (4.51) does not hold.

Proposition 4.5. *If $x_0 \in M \cap N(T)$, then*

$$\omega(M, \delta) \geq \|x_0\|.$$

Proof.   Clearly, if $x_0 \in M \cap N(T)$, then $\|Tx_0\| = 0 \leq \delta$ for every $\delta > 0$. Hence $\omega(M, \delta) \geq \|x_0\|$. $\square$

Question: For all injective operators T, can we expect to have convergence in (4.51) for any $M \subseteq X$?

The answer, in general, is not in affirmative. To see this, let us consider the set

$$M_\rho := \{x : \|x\| \leq \rho\}.$$

Clearly, $\omega(M_\rho, \delta) \leq \rho$. But we have the following discouraging situation.

Theorem 4.27. *If T is not bounded below and $M_\rho := \{x : \|x\| \leq \rho\}$, then*

$$\omega(M_\rho, \delta) = \rho.$$

Proof.   Suppose T is not bounded below. Then there exists a sequence (x_n) in X such that $\|x_n\| = 1$ for all $n \in \mathbb{N}$ and $\|Tx_n\| \to 0$ as $n \to \infty$. Now, for any $\rho > 0$, let $u_n := \rho x_n$, $n \in \mathbb{N}$. Then $u_n \in M_\rho$, and $\|Tu_n\| \to 0$ as $n \to \infty$ so that there exists $N \in \mathbb{N}$ such that $\|Tu_N\| \leq \delta$. Thus,

$$\rho = \|u_N\| \leq \omega(M_\rho, \delta) \leq \rho.$$

Thus, $\omega(M_\rho, \delta) = \rho$. $\square$

Let us recall the following:

- If $R(T)$ is not closed, then T is not bounded below.

- If T is a compact of infinite rank, then $R(T)$ is not closed.

Thus, for every operator with $R(T)$ not closed, in particular, for every compact operator of infinite rank, the set $M_\rho := \{x : \|x\| \leq \rho\}$ is not suitable as a source set to measure the order of convergence of a regularization method.

4.5.1   *Estimates for $\omega(M, \delta)$*

In the last subsection we have seen that if T is not bounded below, then the set $M_\rho := \{x : \|x\| \leq \rho\}$ is not a suitable source set for considering reconstruction algorithms. In the following we specify two commonly used source sets for the reconstructions corresponding to the source conditions specified in Theorem 4.12.

For $\rho > 0$, let

$$M_1 := \{x = T^*u : \|u\| \leq \rho\}$$
$$M_2 := \{x = T^*Tu : \|u\| \leq \rho\}.$$

Theorem 4.28. *The following inequalities hold:*

(i) $\omega(M_1, \delta) \le \rho^{1/2}\delta^{1/2}$,

(ii) $\omega(M_2, \delta) \le \rho^{1/3}\delta^{2/3}$.

Proof. Let $x \in M_1$ be such that $\|Tx\| \le \delta$. By the definition of M_1, there exists $u \in X$ such that $x = T^*u$ with $\|u\| \le \rho$. Then we have

$$\|x\|^2 = \langle x, x \rangle = \langle x, T^*u \rangle = \langle Tx, u \rangle \le \|Tx\|\,\|u\| \le \delta\,\rho.$$

Hence, we have $\omega(M_1, \delta) \le \rho^{1/2}\delta^{1/2}$.

Next, let $x \in M_2$ such that $\|Tx\| \le \delta$. Let $u \in X$ be such that $x = T^*Tu$ with $\|u\| \le \rho$. Then we have

$$\|x\|^2 = \langle x, x \rangle = \langle x, T^*Tu \rangle = \langle Tx, Tu \rangle \le \|Tx\|\,\|Tu\| \le \delta\,\|Tu\|.$$

Note that

$$\langle Tu, Tu \rangle = \langle u, T^*Tu \rangle = \langle u, x \rangle \le \|u\|\,\|x\| \le \rho\|x\|.$$

Hence, we get

$$\|x\|^2 \le \delta\,\|Tu\| \le \delta(\rho\|x\|)^{1/2}.$$

From this we get $\|x\| \le \rho^{1/3}\delta^{2/3}$. Thus, $\omega(M_2, \delta) \le \rho^{1/3}\delta^{2/3}$. $\qquad\square$

In view of Remark 4.8, the source sets M_1 and M_2 considered above are special cases of

$$M_{\nu,\rho} = \{x = (T^*T)^\nu u : \|u\| \le \rho\}, \quad \nu > 0.$$

Corresponding to the set $M_{\nu,\rho}$ we have the following result.

Theorem 4.29. *The following inequality hold:*

$$\omega(M_{\nu,\rho}, \delta) \le \rho^{1/(2\nu+1)}\delta^{2\nu/(2\nu+1)}.$$

Proof. Let $x = (T^*T)^\nu u$ for some $u \in X$ and $\nu > 0$. We observe that

$$x = (T^*T)^\nu u = [(T^*T)^{\frac{2\nu+1}{2}}]^{\frac{2\nu}{2\nu+1}} u = A^\tau u,$$

where $A = (T^*T)^{(2\nu+1)/2}$ and $\tau = 2\nu/(2\nu + 1)$. Hence, from Lemma 4.5 we obtain

$$\|x\| = \|A^\tau u\| \le \|Au\|^\tau\|u\|^{1-\tau}.$$

But,

$$\|Au\| = \|(T^*T)^{\frac{2\nu+1}{2}}u\| = \|(T^*T)^{\frac{1}{2}}x\| = \|Tx\|.$$

Hence,

$$\|x\| \le \|Tx\|^{\frac{2\nu}{2\nu+1}}\|u\|^{\frac{1}{2\nu+1}}.$$

Thus, for every $x \in M_{\nu,\rho}$ with $\|Tx\| \le \delta$, we obtain

$$\|x\| \le \delta^{\frac{2\nu}{2\nu+1}}\rho^{\frac{1}{2\nu+1}}$$

so that the required estimate follows. $\qquad\square$

Remark 4.17. Obviously, results in Theorem 4.28 are particular cases of Theorem 4.29 obtained by taking $\nu = 1/2$ and $\nu = 1$, respectively. However, the proof we gave for Theorem 4.28 for these special cases is simpler than that of Theorem 4.29, as the latter made use of Lemma 4.5. $\Diamond$

The following theorem shows that the estimate in Theorem 4.29 is sharp.

Theorem 4.30. *Let T be a compact operator with singular system $\{(\sigma_n, u_n, v_n) : n \in \mathbb{N}\}$. Then there exists a sequence (δ_n) of positive real numbers such that*

$$\omega(M_{\nu,\rho}, \delta_n) \geq \rho^{1/(\nu+1)} \delta_n^{\nu/(\nu+1)} \quad \forall n \in \mathbb{N}.$$

Proof. For $n \in \mathbb{N}$, let $x_n := \rho\sigma_n^{2\nu} u_n$ and $\delta_n := \rho\sigma_n^{2\nu+1}$. Then we have

$$x_n := \rho\sigma_n^{2\nu} u_n = (T^*T)^\nu(\rho u_n) = (T^*T)^\nu w_n$$

with $w_n := \rho u_n$. Note that $\|w_n\| = \|\rho u_n\| = \rho$ and

$$\|Tx_n\| = \rho\sigma_n^{2\nu}\|Tu_n\| = \rho\sigma_n^{2\nu+1} = \delta_n.$$

Hence, $x_n \in M_{\nu,\rho}$ and $\omega(M_{\nu,\rho}, \delta_n) \geq \|x_n\|$. But,

$$\|x_n\| \doteq \rho\sigma_n^{2\nu} = \rho\left(\frac{\delta_n}{\rho}\right)^{\frac{2\nu}{2\nu+1}} = \rho^{\frac{1}{2\nu+1}} \delta_n^{\frac{2\nu}{2\nu+1}}.$$

Thus, we have proved that $\omega(M_{\nu,\rho}, \delta_n) \geq \rho^{1/(2\nu+1)} \delta_n^{2\nu/(2\nu+1)}$. $\square$

In view of Theorems 4.29 and 4.30, a regularization method R for equation (4.1) is said to be of **order optimal** with respect to the source set $M_{\nu,\rho} := \{x = (T^*T)^{\nu/2} u : \|u\| \leq \rho\}$ if

$$\|\hat{x} - Ry^\delta\| \leq c_0\rho^{1/2\nu+1} \delta^{2\nu/(2\nu+1)}$$

whenever $y^\delta \in Y$ is such that $\|y - y^\delta\| \leq \delta$.

4.5.2   *Illustration with differentiation problem*

Let us consider the problem of differentiation, namely the problem of finding derivative of a function. Thus, given a function $y : [0,1] \to \mathbb{R}$ which is differentiable almost everywhere on $[0,1]$, we would like to find an integrable function $x : [0,1] \to \mathbb{R}$ such that $y' = x$ a.e. This problem is equivalent to solving the integral equation

$$(Tx)(t) := \int_0^t x(s)ds = y(s), \quad 0 \leq s \leq 1.$$

To make our discussion simple, let us assume that the function x we are looking for is differentiable and $x'(1) = 0$. Then we have

$$\|x\|_2^2 = \int_0^1 x(t)x(t)dt$$

$$= \left[x(t) \int_0^t x(t)ds \right]_0^1 - \int_0^1 x'(t) \left(\int_0^t x(s)ds \right) dt$$

$$= - \int_0^1 x'(t) \left(\int_0^t x(s)ds \right) dt$$

$$\leq \|x'\|_2 \|Tx\|_2.$$

The above inequality shows that if $\|x'\|_2 \leq \rho$ and $\|Tx\|_2 \leq \delta$, then we have $\|x\|_2 \leq \rho^{1/2}\delta^{1/2}$. Thus, if we take

$$M = \{x \in C^1[0,1] : x(1) = 0, \|x'\|_2 \leq \rho\},$$

then we get

$$\omega(M,\delta) \leq \rho^{1/2}\delta^{1/2}.$$

More generally, if we define

$$H^1(0,1) := \{x \in L^2(0,1) : x \text{ absolutely continuous}\}$$

and

$$M_\rho = \{x \in H^1(0,1) : x(1) = 0, \|x'\|_2 \leq \rho\},$$

then we get

$$\omega(M_\rho,\delta) \leq \rho^{1/2}\delta^{1/2}.$$

We may observe, in this case, that

$$(T^*u)(t) = \int_t^1 u(s)ds, \quad 0 \leq t \leq 1,$$

so that

$$R(T^*) = \{x \in H^1(0,1) : x(1) = 0\}$$

and

$$M_\rho = \{x = T^*u : \|u\|_2 \leq \rho\}.$$

Thus, the estimate $\omega(M_\rho,\delta) \leq \rho^{1/2}\delta^{1/2}$ is sharp for the set M_ρ.

4.5.3   *Illustration with backward heat equation*

Consider ill-posed problem associated with the backward heat conduction problem discussed in Sections 4.1.2 and 4.4.3, where the problem is to determine the temperature $f := u(\cdot, t_0)$ at a time $t = t_0$ of a thin wire of length ℓ using the knowledge of the temperature $g := u(\cdot, \tau)$ at a later time $t = \tau > t_0$. Recall that, this problem is equivalent to that of solving the compact operator equation

$$Kf = g,$$

with K is defined by

$$K\varphi := \sum_{n=1}^{\infty} \sigma_n \langle \varphi, \varphi_n \rangle \varphi_n, \quad \varphi \in L^2[0, \ell],$$

where

$$\sigma_n := e^{-\lambda_n^2(\tau - t_0)}, \quad \varphi_n(s) := \sqrt{(2/\ell)} \, \sin(\lambda_n s)$$

with $\lambda_n := cn\pi/\ell$ for $s \in [0, \ell]$ and $n \in \mathbb{N}$.

Now, using Hölder's inequality, for $p \in (1, \infty)$ and q satisfying $1/p + 1/q = 1$,

$$\|f\|_2^2 = \sum_{n=1}^{\infty} e^{-2\lambda_n^2 t_0} |\langle f_0, \varphi_n \rangle|^2$$

$$= \sum_{n=1}^{\infty} |\langle f_0, \varphi_n \rangle|^{2/p} |\langle f_0, \varphi_n \rangle|^{2/q} e^{-2\lambda_n^2 t_0}$$

$$\leq \Big(\sum_{n=1}^{\infty} |\langle f_0, \varphi_n \rangle|^2 \Big)^{1/p} \Big(\sum_{n=1}^{\infty} |\langle f_0, \varphi_n \rangle|^2 e^{-2\lambda_n^2 t_0 q} \Big)^{1/q}.$$

Taking $q = \tau/t_0$ we have $p = 1/(1 - t_0/\tau)$ and noting that $g = Tf$, we have

$$\|f\|_2^2 \leq \Big(\sum_{n=1}^{\infty} |\langle f_0, \varphi_n \rangle|^2 \Big)^{(1 - t_0/\tau)} \Big(\sum_{n=1}^{\infty} |\langle f_0, \varphi_n \rangle|^2 e^{-2\lambda_n^2 \tau} \Big)^{t_0/\tau}$$

$$= \|f_0\|_2^{2(1 - t_0/\tau)} \|Kf\|_2^{2t_0/\tau}.$$

Consequently, taking

$$M_\rho := \{ f \in L^2[0, \ell] : f = K_{t_0}\varphi, \ \|\varphi\|_2 \leq \rho \} \tag{4.52}$$

we have

$$\omega(M_\rho, \delta) \leq \rho^{1 - t_0/\tau} \delta^{t_0/\tau}. \tag{4.53}$$

Here K_{t_0} is the operator defined as in Remark 4.1.

4.6 General Source Conditions

4.6.1 *Why do we need a general source condition?*

Suppose X and Y are Hilbert spaces and $T : X \to Y$ is a compact operator with singular system $\{(\sigma_n; u_n, v_n) : n \in \mathbb{N}\}$. Then we know (see Remark 4.7) that for $\nu > 0$, $x \in R((T^*T)^\nu)$ if and only if

$$\sum_{n=1}^{\infty} \frac{|\langle \hat{x}, u_n \rangle|^2}{\sigma_n^{4\nu}} < \infty. \tag{4.54}$$

In Section 4.4.3 we have seen that, in the case of backward heat conduction problem, if the problem is to determine the temperature at a time $t = t_0 > 0$ from the knowledge of the temperature at time $t = \tau > t_0$, the condition (4.54) is satisfied with $\nu = t_0/[2(\tau - t_0)]$, whereas, for the case of $t_0 = 0$, then the condition (4.54) has to be additionally imposed on the initial temperature $f_0 := u(\cdot, 0)$. In this case, condition (4.54) takes the form

$$\sum_{n=1}^{\infty} e^{4\nu\lambda_n^2\tau} |\langle f_0, \varphi_n \rangle|^2 < \infty,$$

where $\lambda_n = c\pi n/\ell$ with $c > 0$. The above considerations motivate us to have a source condition which is milder than the one in (4.54). In this regard, one may consider a condition of the form

$$\sum_{n=1}^{\infty} \frac{|\langle \hat{x}, u_n \rangle|^2}{\varphi(\sigma_n^2)} < \infty, \tag{4.55}$$

where the decay rate of the sequence $(\varphi(\sigma_n^2))$ is much more slowly than $(\sigma_n^{4\nu})$ for any $\nu > 0$. Note that

$$\sum_{n=1}^{\infty} \frac{|\langle \hat{x}, u_n \rangle|^2}{\varphi(\sigma_n^2)} < \infty \iff \hat{x} \in R([\varphi(T^*T)]^{1/2}). \tag{4.56}$$

Motivated by the above, we consider a source condition of the form

$$\hat{x} \in R([\varphi(T^*T)]^{1/2}).$$

for any bounded operator $T : X \to Y$, where φ is a continuous, strictly monotonically increasing real valued function defined on an interval $[0, a]$ containing the spectrum of T^*T such that $\lim_{\lambda \to 0} \varphi(\lambda) = 0$. We call such a function a **source function** for the operator T on $[0, a]$, where $a \geq \|T\|^2$.

Remark 4.18. Since the function $\lambda \mapsto \sqrt{\phi(\lambda)}$ is continuous on $[0, a]$, the definition of the operator $[\phi(T^*T)]^{1/2}$ is as discussed in Section 2.3.4 or in Section 2.3.7. Our preference to the map $\lambda \mapsto \sqrt{\phi(\lambda)}$ instead of $\lambda \mapsto \phi(\lambda)$ is due to the equivalence (4.56). This preference also facilitates to make some of the calculations involved in the following analysis simpler. $\diamond$

Clearly, for each $\nu > 0$, the function $\varphi(\lambda) = \lambda^\nu$, $\lambda > 0$, defines a source function for every bounded operator T. The following consideration will generate another class of source functions.

Suppose $T \in \mathcal{K}(X, Y)$ with singular system $(\sigma_n; u_n, v_n)$, $n \in \mathbb{N}$, and suppose φ is a source function for T. Now, suppose that (σ_n) decays *exponentially*, say σ_n is of the form

$$\sigma_n = e^{-\gamma n^p}, \quad n \in \mathbb{N},$$

for some $\beta > 0$ and $p > 0$. Then for any $q > 0$, we have

$$n^q = \left[\frac{1}{2\gamma} \ln \left(\frac{1}{\sigma_n^2} \right) \right]^{q/p}.$$

In this case the condition (4.54) takes the form

$$\sum_{n=1}^{\infty} e^{4\nu\gamma n^p} |\langle x, u_n \rangle|^2 < \infty. \tag{4.57}$$

· But, if we define

$$\varphi(\lambda) := \left[\frac{1}{2\gamma} \ln \left(\frac{1}{\lambda} \right) \right]^{-q/p}, \quad \lambda \in [0, a], \tag{4.58}$$

then we see that the condition (4.55) takes the form

$$\sum_{n=1}^{\infty} n^q |\langle x, u_n \rangle|^2 < \infty. \tag{4.59}$$

Clearly, the condition (4.59) is much weaker than the condition (4.57).

Example 4.3. In the case of backward heat conduction problem (cf. Section 4.1.2) of determining the initial temperature $f_0 := u(\cdot, 0)$ from the knowledge of the temperature at $t = \tau$, the source set M_ρ in (4.52) is of no use. In this case, we have to rely on a general source condition of the form (4.55). In this case, we have

$$\sigma_n := e^{-\lambda_n^2 \tau}, \quad \lambda_n := cn\pi/\ell, \quad n \in \mathbb{N}.$$

Thus, $p = 2$ and $\gamma = \tau c^2 \pi^2 / \ell^2$ so that φ in (4.58) takes the form

$$\varphi(\lambda) := \left[\frac{\ell^2}{2\tau c^2 \pi^2} \ln \left(\frac{1}{\lambda} \right) \right]^{-q/2}, \quad \lambda \in [0, a],$$

and

$$f_0 \in R([\varphi(K_\tau^* K_\tau)]^{1/2}) \iff \sum_{n=1}^{\infty} n^q |\langle f, \varphi_n \rangle|^2 < \infty,$$

where K_τ and φ_n are as in Remark 4.1.   ◇

4.6.2 Error estimates for Tikhonov regularization

Recall that in *Tikhonov regularization*, we have the regularization family $\{R_\alpha : \alpha > 0\}$ with

$$R_\alpha := (T^*T + \alpha I)^{-1}T^*, \quad \alpha > 0.$$

In this case we know that if $y \in D(T^\dagger)$ and for $\delta > 0$, if $y^\delta \in Y$ satisfies $\|y - y^\delta\| \le \delta$, then

$$\|\hat{x} - x_\alpha\| \to 0 \quad \text{as} \quad \alpha \to 0$$

and

$$\|x_\alpha - x_\alpha^\delta\| \le \frac{\delta}{\sqrt{\alpha}},$$

where $\hat{x} := T^\dagger y$, $x_\alpha := R_\alpha y$ and $x_\alpha^\delta := R_\alpha y^\delta$. We may also recall that

$$\hat{x} - x_\alpha = \alpha(T^*T + \alpha I)^{-1}\hat{x}.$$

In the following we obtain error estimates under the assumption that $\hat{x} \in R([\varphi(T^*T)]^{1/2})$ for some source function $\varphi : [0, a] \to \mathbb{R}$.

Theorem 4.31. *Suppose* $\hat{x} = [\varphi(T^*T)]^{1/2}u$ *for some* $u \in X$ *and for some source function* φ *satisfying*

$$\sup_{0 < \lambda \le a} \frac{\alpha\sqrt{\varphi(\lambda)}}{\lambda + \alpha} \le \sqrt{\varphi(\alpha)}, \tag{4.60}$$

where $a \ge \|T\|^2$. *Then*

$$\|\hat{x} - x_\alpha\| \le \|u\|\sqrt{\varphi(\alpha)}.$$

Proof. First we observe that

$$\|\hat{x} - x_\alpha\| = \|\alpha(T^*T + \alpha I)^{-1}[\varphi(T^*T)]^{1/2}u\|$$

$$\le \|u\| \sup_{0 < \lambda \le a} \frac{\alpha\sqrt{\varphi(\lambda)}}{\lambda + \alpha}.$$

Hence, the required result follows from the inequality (4.60). $\qquad\square$

The following lemma specifies a condition under which the condition (4.60) holds.

Lemma 4.8. *Suppose* φ *is a source function for* T *such that* $\lambda \mapsto \sqrt{\varphi(\lambda)}$ *is concave. Then the inequality (4.60) is satisfied.*

Proof.   Since the map $\lambda \mapsto \sqrt{\varphi(\lambda)}$ is concave and monotonically increasing, we have

$$\frac{\alpha\sqrt{\varphi(\lambda)}}{\lambda + \alpha} \leq \sqrt{\varphi\left(\frac{\alpha\lambda}{\lambda + \alpha}\right)} \leq \sqrt{\varphi(\alpha)}.$$

This completes the proof.   □

Remark 4.19. It can be seen that the functions

$$\lambda \mapsto \lambda^{2\nu}$$

for $0 < \nu \leq 1/2$, and the function

$$\lambda \mapsto \left[\frac{1}{2\gamma} \ln\left(\frac{1}{\lambda}\right)\right]^{-q/p}$$

for $p > 0, q > 0, \gamma > 0$ are concave.   ◊

Now, we derive estimates for the error $\|\hat{x} - x_\alpha^\delta\|$ under certain a priori and a posteriori parameter choice strategies. For this purpose we shall use the source set

$$M_{\varphi,\rho} := \{x \in X : x = [\varphi(T^*T)]^{1/2}u, \|u\| \leq \rho\}$$

for some $\rho > 0$. We shall also make use of the function ψ defined by

$$\psi(\lambda) = \lambda\varphi^{-1}(\lambda), \quad \lambda \in [0, a\varphi(a)].$$

4.6.3   *Parameter choice strategies*

(i) An a priori parameter choice

From the definition of the function ψ, the relations in the following lemma can be derived easily.

Lemma 4.9. *For $\delta > 0$ and $\alpha > 0$,*

$$\rho^2\alpha\varphi(\alpha) = \delta^2 \iff \alpha = \varphi^{-1}\left[\psi^{-1}\left(\frac{\delta^2}{\rho^2}\right)\right],$$

and in that case

$$\rho^2\varphi(\alpha) = \frac{\delta^2}{\alpha} = \rho^2\psi^{-1}\left(\frac{\delta^2}{\rho^2}\right).$$

Theorem 4.32. *Suppose* $\hat{x} \in M_{\varphi,\rho}$, *where* φ *satisfies the condition (4.60).* *For* $\delta > 0$, *let* α *be chosen such that* $\rho^2 \alpha \varphi(\alpha) = \delta^2$. *Then*

$$\|\hat{x} - x_\alpha^\delta\| \leq 2\rho \sqrt{\psi^{-1}\left(\frac{\delta^2}{\rho^2}\right)}.$$

Proof. By Theorem 4.31 and using the estimate for $\|x_\alpha - x_\alpha^\delta\|$ we have

$$\|\hat{x} - x_\alpha^\delta\| \leq \rho\sqrt{\varphi(\alpha)} + \frac{\delta}{\sqrt{\alpha}}.$$

Thus, the estimate follows from Lemma 4.9. $\qquad\square$

Remark 4.20. (a) Suppose $\varphi(\lambda) = \lambda^{2\nu}$ for $\nu > 0$. Then we see that

$$\varphi^{-1}(\lambda) = \lambda^{1/2\nu}, \quad \psi(\lambda) = \lambda^{(2\nu+1)/2\nu}, \quad \psi^{-1}(\lambda) = \lambda^{2\nu/(2\nu+1)}.$$

Hence, in this case, we have

$$\rho\sqrt{\psi^{-1}\left(\frac{\delta^2}{\rho^2}\right)} = \rho^{1/(2\nu+1)}\delta^{2\nu/(2\nu+1)}.$$

(b) For $\lambda \in [0, a]$, let $\varphi(\lambda) := \left[\frac{1}{2\gamma}\ln\left(\frac{1}{\lambda}\right)\right]^{-q/p}$. We note that

$$\varphi(\lambda) = s \iff \lambda = e^{-2\gamma/s^{q/p}}.$$

Hence,

$$\psi(s) = s\varphi^{-1}(s) = se^{-2\gamma/s^{q/p}}.$$

Now,

$$\psi(s) = t \iff \frac{1}{t} = \frac{1}{s}e^{2\gamma/s^{q/p}} \iff \ln\frac{1}{t} = \ln\frac{1}{s} + \frac{2\gamma}{s^{q/p}}.$$

Hence,

$$s = \left[\frac{2\gamma}{\ln\frac{1}{t} - \ln\frac{1}{s}}\right]^{p/q} = \left(\frac{2\gamma}{\ln\frac{1}{t}}\right)^{p/q}\left[1 - \frac{\ln\frac{1}{s}}{\ln\frac{1}{t}}\right]^{-p/q}.$$

But, we have

$$\frac{\ln\frac{1}{s}}{\ln\frac{1}{t}} = \frac{\ln\frac{1}{s}}{\ln\frac{1}{s} + 2\gamma/s^{q/p}} \to 0 \quad\text{as}\quad s \to 0.$$

Hence

$$\psi^{-1}(t) = (2\gamma)^{p/q}\left[\ln\frac{1}{t}\right]^{-p/q}[1 + o(1)]$$

so that

$$\rho\sqrt{\psi^{-1}\left(\frac{\delta^2}{\rho^2}\right)} = (2\gamma)^{p/2q}\left[\ln\frac{\rho}{\delta}\right]^{-p/q}[1 + o(1)].$$

$\Diamond$

In view of the above remark we obtain the following corollaries.

Corollary 4.9. *Suppose $\hat{x} \in M_{\varphi,\rho}$ with $\varphi(\lambda) := \lambda^{2\nu}$, $0 \leq \lambda \leq a$, for some $\nu \in (0, 1/2]$, and for $\delta > 0$, let α be chosen such that $\alpha\varphi(\alpha) = \delta^2/\rho^2$. Then*

$$\|\hat{x} - x_\alpha^\delta\| \leq 2\rho^{1/(2\nu+1)}\delta^{2\nu/(2\nu+1)}.$$

Corollary 4.10. *Suppose $\hat{x} \in M_{\varphi,\rho}$ with*

$$\varphi(\lambda) := \left[\frac{1}{2\gamma}\ln\left(\frac{1}{\lambda}\right)\right]^{-q/p}, \quad \lambda \in [0, a],$$

for some positive real numbers p, q and γ, and let α be chosen such that $\alpha\varphi(\alpha) = \delta^2/\rho^2$. Then

$$\|\hat{x} - x_\alpha^\delta\| \leq (2\gamma)^{p/2q}\left[\ln\frac{\rho}{\delta}\right]^{-p/q}[1 + o(1)].$$

(i) An a posteriori parameter choice

Now we obtain estimate for the error $\|\hat{x} - x_\alpha^\delta\|$ by choosing the parameter α as per Morozov-type discrepancy principle

$$\delta \leq \|Tx_\alpha^\delta - y^\delta\| \leq c\,\delta \tag{4.61}$$

for some $c \geq 1$. We shall make use of the *Jensen's inequality* stated in the following lemma. For its proof, one may refer to Rudin [66].

Lemma 4.10. (Jensen's inequality) *Let μ be a positive measure on a σ-algebra $\mathcal{A}$ on a set Ω such that $\mu(\Omega) = 1$. If f is a real valued function on Ω integrable with respect to μ and if J is an integral containing the range of f, then for every convex function g on J,*

$$g\left(\int_\Omega f\,d\mu\right) \leq \int_\Omega (g \circ f)d\mu.$$

Theorem 4.33. *Suppose $y \in R(T)$ and $\hat{x} \in M_{\varphi,\rho}$, where φ is a concave function. For $\delta > 0$, let α satisfy the discrepancy principle (4.61) for some $c \geq 1$. Then*

$$\|\hat{x} - x_\alpha^\delta\| \leq (c+1)\rho\sqrt{\psi^{-1}\left(\frac{\delta^2}{\rho^2}\right)}.$$

Proof. Since x_α^δ minimizes the function $x \mapsto \|Tx - y^\delta\|^2 + \alpha\|x\|^2$ and $y = T\hat{x}$, we have

$$\begin{aligned}
\delta^2 + \alpha\|x_\alpha^\delta\|^2 &\leq \|Tx_\alpha^\delta - y^\delta\|^2 + \alpha\|x_\alpha^\delta\|^2 \\
&\leq \|T\hat{x} - y^\delta\|^2 + \alpha\|\hat{x}\|^2 \\
&= \|y - y^\delta\|^2 + \alpha\|\hat{x}\|^2 \\
&\leq \delta^2 + \alpha\|\hat{x}\|^2.
\end{aligned}$$

Hence, $\|x_\alpha^\delta\| \le \|\hat{x}\|$. Therefore,

$$
\begin{aligned}
\|\hat{x} - x_\alpha^\delta\|^2 &= \langle \hat{x}, \hat{x} \rangle - 2\,\mathrm{Re}\langle \hat{x}, x_\alpha^\delta \rangle + \langle x_\alpha^\delta, x_\alpha^\delta \rangle \\
&\le 2\left[\langle \hat{x}, \hat{x} \rangle - \mathrm{Re}\langle \hat{x}, x_\alpha^\delta \rangle \right] \\
&= 2\,\mathrm{Re}\langle \hat{x}, \hat{x} - x_\alpha^\delta \rangle \\
&= 2\,|\langle \hat{x}, \hat{x} - x_\alpha^\delta \rangle|.
\end{aligned}
$$

Since $\hat{x} \in M_{\varphi,\rho}$, there exists $u \in X$ such that $\hat{x} = [\varphi(T^*T)]^{1/2}u$ with $\|u\| \le \rho$. Hence,

$$
\begin{aligned}
\langle \hat{x}, \hat{x} - x_\alpha^\delta \rangle &= \langle [\varphi(T^*T)]^{1/2}u, \hat{x} - x_\alpha^\delta \rangle \\
&= \langle u, [\varphi(T^*T)]^{1/2}(\hat{x} - x_\alpha^\delta) \rangle \\
&\le \rho \|[\varphi(T^*T)]^{1/2}(\hat{x} - x_\alpha^\delta)\|.
\end{aligned}
$$

Thus,

$$
\|\hat{x} - x_\alpha^\delta\|^2 \le 2\,\rho \|[\varphi(T^*T)]^{1/2}(\hat{x} - x_\alpha^\delta)\|
$$

so that

$$
\frac{\|\hat{x} - x_\alpha^\delta\|}{2\rho} \le \frac{\|[\varphi(T^*T)]^{1/2}(\hat{x} - x_\alpha^\delta)\|}{\|\hat{x} - x_\alpha^\delta\|}.
$$

Therefore, using spectral theorem for the positive self-adjoint operator T^*T with respect to its spectral family $\{E_\lambda : 0 < \lambda \le a\}$,

$$
\begin{aligned}
\frac{\|\hat{x} - x_\alpha^\delta\|^2}{4\rho^2} &\le \frac{\|[\varphi(T^*T)]^{1/2}(\hat{x} - x_\alpha^\delta)\|^2}{\|\hat{x} - x_\alpha^\delta\|^2} \\
&= \frac{\int_0^a \varphi(\lambda)d\,\|E_\lambda(\hat{x} - x_\alpha^\delta)\|^2}{\int_0^a d\,\|E_\lambda(\hat{x} - \tilde{x}_\alpha)\|^2}.
\end{aligned}
$$

We may observe that the measure μ on $[0, a]$ defined by

$$
d\mu(\lambda) := \frac{d\,\|E_\lambda(\hat{x} - x_\alpha^\delta)\|^2}{\int_0^a d\,\|E_\lambda(\hat{x} - \tilde{x}_\alpha)\|^2}
$$

satisfies $\mu([0, a]) \doteq 1$. Since φ^{-1} is convex, applying φ^{-1} on the above inequality and by using *Jensen's inequality* (see Lemma 4.10), we get

$$
\begin{aligned}
\varphi^{-1}\left[\frac{\|\hat{x} - x_\alpha^\delta\|^2}{4\rho^2} \right] &= \varphi^{-1}\left(\int_0^a \varphi(\lambda)d\mu(\lambda) \right) \\
&\le \int_0^a \lambda\, d\mu(\lambda).
\end{aligned}
$$

But,

$$\int_0^a \lambda d\mu(\lambda) = \frac{\int_0^a \lambda d \, \|E_\lambda(\hat{x} - x_\alpha^\delta)\|^2}{\int_0^a d \, \|E_\lambda(\hat{x} - x_\alpha^\delta)\|^2}$$

$$= \frac{\|T(\hat{x} - x_\alpha^\delta)\|^2}{\|\hat{x} - x_\alpha^\delta\|^2}.$$

Thus, we obtain

$$\varphi^{-1}\left[\frac{\|\hat{x} - x_\alpha^\delta\|^2}{4\rho^2}\right] \leq \frac{\|T(\hat{x} - x_\alpha^\delta)\|^2}{\|\hat{x} - x_\alpha^\delta\|^2}.$$

But

$$\|T(\hat{x} - x_\alpha^\delta)\| = \|y - Tx_\alpha^\delta\| \leq \|y - y^\delta\| + \|y^\delta - Tx_\alpha^\delta\| \leq (1 + c)\,\delta.$$

Hence

$$\varphi^{-1}\left[\frac{\|\hat{x} - x_\alpha^\delta\|^2}{(c+1)^2 \rho^2}\right] \leq \varphi^{-1}\left[\frac{\|\hat{x} - x_\alpha^\delta\|^2}{4\rho^2}\right]$$

$$\leq \frac{\|T(\hat{x} - x_\alpha^\delta)\|^2}{\|\hat{x} - x_\alpha^\delta\|^2}$$

$$\leq \frac{(c+1)^2 \delta^2}{\|\hat{x} - x_\alpha^\delta\|^2}.$$

Since $\psi(\lambda) = \lambda \varphi^{-1}(\lambda)$, it follows from the above inequality that

$$\psi\left(\frac{\|\hat{x} - x_\alpha^\delta\|^2}{(c+1)^2 \rho^2}\right) = \frac{\|\hat{x} - x_\alpha^\delta\|^2}{(c+1)^2 \rho^2} \, \varphi^{-1}\left[\frac{\|\hat{x} - x_\alpha^\delta\|^2}{(c+1)^2 \rho^2}\right]$$

$$\leq \frac{\delta^2}{\rho^2}.$$

Thus,

$$\|\hat{x} - x_\alpha^\delta\| \leq (c+1)\,\rho\sqrt{\psi^{-1}\left(\frac{\delta^2}{\rho^2}\right)}$$

which completes the proof. $\qquad\qquad\qquad\qquad\qquad\qquad\qquad\qquad \Box$

In view of Remark 4.20 we obtain the following corollaries.

Corollary 4.11. *Suppose* $y \in R(T)$ *and* $\hat{x} \in M_{\varphi,\rho}$ *with* $\varphi(\lambda) := \lambda^{2\nu}$, $0 \leq \lambda \leq a$, *for some* $\nu \in (0, 1/2]$, *and for* $\delta > 0$, *let* α *satisfy the discrepancy principle (4.61) for some* $c \geq 1$. *Then*
$$\|\hat{x} - x_\alpha^\delta\| \leq (c+1)\rho^{1/(2\nu+1)}\delta^{2\nu/(2\nu+1)}.$$

Corollary 4.12. *Suppose* $y \in R(T)$ *and* $\hat{x} \in M_{\varphi,\rho}$ *with*
$$\varphi(\lambda) := \left[\frac{1}{2\gamma}\ln\left(\frac{1}{\lambda}\right)\right]^{-q/p}, \quad \lambda \in [0, a],$$
for some positive real numbers p, q *and* γ, *and let* α *satisfy the discrepancy principle (4.61) for some* $c \geq 1$. *Then*
$$\|\hat{x} - x_\alpha^\delta\| \leq (c+1)(2\gamma)^{p/2q}\left[\ln\frac{\rho}{\delta}\right]^{-p/q}[1 + o(1)].$$

4.6.4 *Estimate for* $\omega(M_{\varphi,\rho}, \delta)$

Now we show that the estimates obtained in Theorems 4.32 and 4.33 are order optimal, by obtaining a sharp estimate for the quantity $\omega(M_{\varphi,\rho}, \delta)$.

Theorem 4.34. *Suppose φ is a source function for T such that φ is concave. Then*

$$\omega(M_{\varphi,\rho}, \delta) \le \rho\sqrt{\psi^{-1}\left(\frac{\delta^2}{\rho^2}\right)},$$

where $\psi(\lambda) := \lambda\varphi^{-1}(\lambda)$ for $\lambda \in [0, a\varphi^{-1}(\lambda)]$.

Proof.   Let $x = [\varphi(T^*T)]^{1/2}u$ with $\|u\| \le \rho$ and $\|Tx\| \le \delta$. Then we have

$$\begin{aligned}
\|x\|^2 &= \langle[\varphi(T^*T)]^{1/2}u, x\rangle \\
&= \langle u, [\varphi(T^*T)]^{1/2}x\rangle \\
&\le \rho\|[\varphi(T^*T)]^{1/2}x\|.
\end{aligned}$$

Thus,

$$\frac{\|x\|}{\rho} \le \frac{\|[\varphi(T^*T)]^{1/2}x\|}{\|x\|}.$$

Hence, using spectral theorem for the positive self-adjoint operator T^*T with respect to its spectral family $\{E_\lambda : 0 < \lambda \le a\}$, we have

$$\begin{aligned}
\frac{\|x\|^2}{\rho^2} &\le \frac{\|\varphi(T^*T)]^{1/2}x\|^2}{\|x\|^2} \\
&= \frac{\int_0^b \varphi(\lambda)d\langle E_\lambda x, x\rangle}{\int_0^b d\langle E_\lambda x, x\rangle}.
\end{aligned}$$

Since φ^{-1} is convex, we can apply Jensen's inequality so that

$$\begin{aligned}
\varphi^{-1}\left(\frac{\|x\|^2}{\rho^2}\right) &\le \frac{\int_0^b \lambda d\langle E_\lambda x, x\rangle}{\int_0^b d\langle E_\lambda x, x\rangle} \\
&= \frac{\|Tx\|^2}{\|x\|^2} \\
&\le \frac{\delta^2}{\|x\|^2}.
\end{aligned}$$

Thus,

$$\psi\left(\frac{\|x\|^2}{\rho^2}\right) = \frac{\|x\|^2}{\rho^2}\varphi^{-1}\left(\frac{\|x\|^2}{\rho^2}\right) \le \frac{\delta^2}{\rho^2}.$$

From this, we obtain

$$\|x\|^2 \le \rho^2\psi^{-1}\left(\frac{\delta^2}{\rho^2}\right)$$

which completes the proof. $\qquad\square$

In view of Remark 4.20, Theorem 4.34 provides the following corollaries.

Corollary 4.13. *If $\varphi(\lambda) := \lambda^{2\nu}$, $0 \le \lambda \le a$, for some $\nu \in (0, 1/2]$, then*

$$\omega(M_{\varphi,\rho}) \le \rho^{1/(2\nu+1)} \delta^{2\nu/(2\nu+1)}.$$

Corollary 4.14. *If*

$$\varphi(\lambda) := \left[\frac{1}{2\gamma} \ln \left(\frac{1}{\lambda} \right) \right]^{-q/p}, \quad \lambda \in [0, a],$$

for some positive real numbers p and q, then

$$\omega(M_{\varphi,\rho}) \le (2\gamma)^{p/2q} \left[\ln \frac{\rho}{\delta} \right]^{-p/q} [1 + o(1)].$$

The following theorem shows that the estimate obtained in Theorem 4.34 is, in fact, sharp.

Theorem 4.35. *Let T be a compact operator with singular values σ_n, $n \in \mathbb{N}$. Then with $\delta_n = \rho \sigma_n [\varphi(\sigma_n^2)]^{1/2}$,*

$$\omega(M_{\varphi,\rho}, \delta_n) \ge \rho \sqrt{\psi^{-1}\left(\frac{\delta_n^2}{\rho^2} \right)}.$$

Proof. Let $w_n = \rho u_n$ so that $\|w_n\| = \rho$ and

$$x_n := [\varphi(T^*T)]^{1/2} w_n = \rho[\varphi(\sigma_n^2)]^{1/2} u_n$$

satisfies

$$w_n \in M_{\varphi,\rho} \quad \text{and} \quad \|Tx_n\| = \rho \sigma_n [\varphi(\sigma_n^2)]^{1/2}.$$

Thus taking $\delta_n := \rho \sigma_n [\varphi(\sigma_n^2)]^{1/2}$, it follows that

$$\omega(M_{\varphi,\rho}, \delta_n) \ge \|x_n\| = \rho[\varphi(\sigma_n^2)]^{1/2}.$$

Now, by Lemma 4.9, we have $\varphi(\sigma_n^2) = \psi^{-1}\left(\delta_n^2 / \rho^2 \right)$. Hence,

$$\omega(M_{\varphi,\rho}, \delta_n) \ge \rho[\varphi(\sigma_n^2)]^{1/2}$$
$$= \rho \sqrt{\psi^{-1}\left(\frac{\delta_n^2}{\rho^2} \right)}.$$

This completes the proof. $\square$

PROBLEMS

In the following, the operators are between Hilbert spaces.

(1) Show that a compact operator of infinite rank is not bounded below.
(2) Let $T : L^2[0,1] \to L^2[0,1]$ be defined by

$$(Tx)(t) = \int_0^t x(t)\,dt, \quad x \in L^2[0,1],\ t \in [0,1].$$

Find $y \in L^2[0,1]$ such that $Tx = y$ does not have an LRN solution.
(3) If $R(T)$ is dense, then $D(T^\dagger) = R(T)$ - Why?
(4) If $R(T)$ is dense, then show that for every $y \in Y$, $\|Tx_\alpha - y\| \to 0$ as $\alpha \to 0$, where $x_\alpha := (T^*T + \alpha I)^{-1}T^*y$.
(5) Suppose $T \in \mathcal{B}(X,Y)$ and $L : D(L) \subseteq X \to X$ is a closed operator with its domain X_0 dense in X such that

$$\|Tx\|^2 + \|x\|^2 \geq \gamma\|x\|^2 \quad \forall x \in D(L).$$

Then prove the following:
(i) The map $(x,u) \mapsto \langle x, u \rangle_0 := \langle Tx, Tu \rangle + \langle x, u \rangle$ defines an inner product on $D(L)$, and $X_0 := D(L)$ with $\langle \cdot, \cdot \rangle_0$ is a Hilbert space.
(ii) $T_0 := T|_{X_0} : X_0 \to Y$ is a bounded linear operator.
(iii) $D(T_0^\dagger) = R(T_0) + R(T)^\perp$, and for $y \in D(T_0^\dagger)$, $x_L := T_0^\dagger y$ is the unique element in

$$\{x \in X_0 : \|Tx - y\| \leq \inf_{u \in X_0} \|Tu - y\|\}$$

such that $\|x_L\| = \inf_{x \in X_0} \|Lx\|$.
(6) Suppose T and L are as in Problem 5. Prove the following:
(i) The operator $T^*T + \alpha L^*L$ is a closed bijective operator from $D(L^*L)$ to X and its inverse $(T^*T + \alpha L^*L)^{-1}$ is a bounded operator.
(ii) For $y \in D(T_0^\dagger)$ and $\alpha > 0$, if $x_{L,\alpha} := (T^*T + \alpha L^*L)^{-1}T^*y$, then $x_{L,\alpha} \to T_0^\dagger y$ as $\alpha \to 0$.
(7) Suppose X is a linear space, Y is a normed linear space and $T : X \to Y$ is a linear operator. Let $y_0 \in Y$, $\delta_0 > 0$ and $V_0 := \{y \in Y : \|y - y_0\| < \delta_0\}$. If the equation $Tx = y$ has a solution for every $y \in V_0$, then show that $Tx = y$ has a solution for every $y \in Y$.
(8) Let X and Y be Banach spaces, X_0 be a subspace of X and $T : X_0 \to Y$ be a closed operator. Show that the map $F : X_0 \to X \times Y$ defined by $F(x) = (x, Tx)$, $x \in X_0$, is a closed operator and $R(F)$ is closed in $X \times Y$.

(9) Let $u \in C[a,b]$ and T be defined by

$$(Tx)(t) = u(t)x(t), \quad x \in L^2[a,b], \ t \in [a,b].$$

Show that $R(T)$ is closed if and only if 0 is an accumulation point of the set $\{|u(t)| : a \le t \le b\}$.

(10) Let X and Y be Hilbert spaces, $T \in \mathcal{B}(X,Y)$ and $A \in \mathcal{B}(X)$. If A is a positive self-adjoint operator and $\mu \ge 0$, then prove

(i) $\|(A + \alpha I)^{-2\mu} A^\mu\| \le (4\alpha)^{-\mu}$, and

(ii) $\|T(T^*T + \alpha I)^{-2\mu-1}(T^*T)^\mu\| \le (4\alpha)^{-\mu-\frac{1}{2}}$.

(11) For $y \in D(T^\dagger)$, prove that

(i) $T^\dagger y \in R(T^*) \iff \sum_{n=1}^\infty |\langle y, v_n\rangle|^2 / \sigma_n^4 < \infty$,

(ii) $T^\dagger y \in R(T^*T) \iff \sum_{n=1}^\infty |\langle y, v_n\rangle|^2 / \sigma_n^6 < \infty$.

(12) Let $\{u_n : n \in \mathbb{N}\}$ be an orthonormal set and for $\nu > 0$ and for a sequence (σ_n) of positive real numbers which converges to 0, let

$$X_\nu := \left\{ x \in X : \sum_{n=1}^\infty \frac{|\langle x, u_n\rangle|^2}{\sigma_n^{4\nu}} < \infty \right\}.$$

Show that

$$\langle x, y\rangle_\nu := \sum_{n=1}^\infty \frac{\langle x, u_n\rangle \langle u_n, y\rangle}{\sigma_n^{4\nu}}$$

defines an inner product on X and X with the inner product $\langle \cdot, \cdot\rangle_\nu$ is a Hilbert space.

(13) Let X_ν be as in Problem 12. Show that for $\nu \le \mu$, $X_\mu \subseteq X_\nu$, and the embedding $X_\mu \hookrightarrow X_\nu$ is a compact operator.

(14) Suppose X_ν, (u_n) and (σ_n) are as in Problem 12. Let φ be a real valued function defined on $\mathrm{cl}\{\sigma_n^2 : n \in \mathbb{N}\}$ such that $\varphi(\sigma_n^2) \to 0$ as $n \to \infty$. Let

$$X_\varphi := \left\{ x \in X : \sum_{n=1}^\infty \frac{|\langle x, u_n\rangle|^2}{\varphi(\sigma_n^2)} < \infty \right\}.$$

Show that

$$\langle x, y\rangle_\varphi := \sum_{n=1}^\infty \frac{\langle x, u_n\rangle \langle u_n, y\rangle}{\varphi(\sigma_n^2)}$$

defines an inner product on X, and X with the inner product $\langle \cdot, \cdot\rangle_\varphi$ is a Hilbert space.

(15) Let $K_0 : L^2[0, \ell] \to L^2[0, \ell]$ defined by

$$K_0 x = \sum_{n=1}^{\infty} e^{-\lambda_n^2 \tau} \langle x, v_n \rangle v_n, \quad \varphi \in L^2[0, \ell],$$

where $\lambda_n = n\pi/\ell$ and $v_n(s) := \sqrt{2/\ell} \sin(\lambda_n s)$ for $s \in [0, \ell]$ and $n \in \mathbb{N}$.
Show that

$$\{x : x'' \in L^2[0, \ell], \ \|x''\|_2 \leq \rho\} = \{x = \varphi(K_0^* K_0) v, \ \|v\|_2 \leq \rho\},$$

where $\varphi(\lambda) := \left[\dfrac{1}{2\tau} \ln \left(\dfrac{1}{\lambda} \right) \right]^{-1}$, $\lambda \in [0, e^{-2\lambda_1^2 \tau}]$.

(16) Using the notations in Section 4.5.3,
    (i) show that $f = (K^* K)^\nu f_0$ with $\nu := t_0/2(\tau - t_0)$, and
    (ii) deduce the estimate $\omega(M_\rho, \delta) \leq \rho^{1 - t_0/\tau} \delta^{t_0/\tau}$ from Theorem 4.29,
    where, $M_\rho := \{f \in L^2[0, \ell] : f = K_{t_0} \varphi, \ \|\varphi\|_2 \leq \rho\}$ with K_{t_0} as in (4.1).

(17) Show that in the backward heat conduction problem (cf. Section 4.1.2)
of determining the initial temperature $f_0 := u(\cdot, 0)$ from the knowledge
of the temperature at $t = \tau$ with $\ell = 1$ and $c = 1$,

$$\{\varphi \in L^2[0, \ell] : \|\varphi''\|_2 \leq \rho\} = \{[\varphi(K_\tau^* K_\tau)]^{1/2} \varphi : \|\varphi\| \leq \rho\},$$

where $\varphi(\lambda) := \left(\dfrac{1}{2\tau} \ln \dfrac{1}{\lambda} \right)^{-1}$.

Chapter 5

Regularized Approximation Methods

5.1 Introduction

As in Chapter 4, we are again interested in obtaining stable approximate solutions for the ill-posed operator equation

$$Tx = y,$$

where $T : X \to Y$ is a bounded linear operator between Hilbert spaces X and Y. Recall that, as approximations for the generalized solution $\hat{x} := T^\dagger y$ of such ill-posed equations whenever $y \in D(T^\dagger)$, we considered regularized solutions

$$x_\alpha^\delta := R_\alpha y^\delta,$$

where y^δ is an approximation to y with

$$\|y - y^\delta\| \leq \delta$$

for some noise level $\delta > 0$, and $\{R_\alpha : \alpha > 0\}$ is a family of bounded operators, known as a *regularization family*. Since R_α is continuous for each $\alpha > 0$, the above regularized solution is often obtained by solving some well-posed problems.

In practice, one may have to satisfy with some approximations $x_{\alpha,n}^\delta$ of x_α^δ obtained by approximating the operators involved in the well-posed problems, say by considering an approximation $R_\alpha^{(n)}$ of R_α, and taking

$$x_{\alpha,n}^\delta := R_\alpha^{(n)} y^\delta.$$

Now, the question is, how well $x_{\alpha,n}^\delta$ approximates $\hat{x}$, by suitable choice of the parameters $\alpha > 0$ and $n \in \mathbb{N}$. Clearly

$$\|\hat{x} - x_{\alpha,n}^\delta\| \leq \|\hat{x} - x_\alpha\| + \|(R_\alpha - R_\alpha^{(n)})y\| + \|R_\alpha^{(n)}\|\delta. \qquad (5.1)$$

In particular, if $\|(R_\alpha - R_\alpha^{(n)})y\| \to 0$ as $n \to \infty$ for each $\alpha > 0$ and if $\alpha := \alpha(\delta, n, y^\delta)$ can be chosen in such a way that $\alpha(\delta, n, y^\delta) \to 0$ and $\|R_\alpha^{(n)}\| \delta \to 0$ as $\delta \to 0$ and $n \to \infty$, then
$$\|\hat{x} - x_{\alpha,n}^\delta\| \to 0 \quad \text{as} \quad \delta \to 0, \, n \to \infty.$$
In such case, we say that $x_{\alpha,n}^\delta$ is a **regularized approximation** of $\hat{x}$. Estimates for the error $\|\hat{x} - x_{\alpha,n}^\delta\|$ can be obtained, once we have estimates for the quantities $\|\hat{x} - x_\alpha\|$, $\|(R_\alpha - R_\alpha^{(n)})y\|$ and $\|R_\alpha^{(n)}\|$.

In this chapter we discuss the above issues in the case of Tikhonov regularization, that is, when
$$R_\alpha := (T^*T + \alpha I)^{-1}T^*, \quad \alpha > 0.$$
Thus, the equation to which an approximate solution is sought is
$$(T^*T + \alpha I)x_\alpha^\delta = T^*y^\delta. \tag{5.2}$$
As approximation of R_α one may consider operators of the form
$$R_\alpha^{(n)} := (A_n + \alpha I)^{-1}B_n, \quad \alpha > 0,$$
for large enough n, say for $n \geq N$, where (A_n) and (B_n) are approximations of T^*T and T^*, respectively, in some sense, so that for each $\alpha > 0$, $A_n + \alpha I$ is bijective for all $n \geq N$. Thus, a regularized approximate solution would be a solution of the equation
$$(A_n + \alpha I)x_{\alpha,n}^\delta = B_n y^\delta. \tag{5.3}$$
If A_n is a finite rank operator then the problem of solving the above equation can be converted into a problem of solving a system of linear algebraic equations. Indeed, for a fixed n, if $R(A_n)$ is spanned by $u_1, \ldots, u_N$, then there exist continuous linear functionals $f_1, \ldots, f_N$ on X such that for every $x \in X$,
$$A_n x = \sum_{j=1}^N f_j(x)u_j,$$
and then $x_{\alpha,n}^\delta$ is given by
$$x_{\alpha,n}^\delta = \frac{1}{\alpha}\Big[B_n y^\delta - \sum_{j=1}^N c_j u_j\Big],$$
where c_j, $j = 1, \ldots, N$, satisfy the system of equations
$$\alpha c_i - \sum_{j=1}^N f_i(u_j)c_j = d_i, \quad j = 1, \ldots, N,$$
with $d_i = f_i(B_n \tilde{y})$, and in that case
$$c_j = f_j(x_{\alpha,n}^\delta), \quad j = 1, \ldots, N.$$

Exercise 5.1. Prove the assertion in the last sentence.

If an approximation (T_n) of T is known, then a choice for A_n and B_n could be

$$A_n := T_n^* T_n, \quad B_n = T_n^*.$$

In this special case, since A_n is positive and self adjoint, equation (5.3) can be solved uniquely. In the next section, we shall be considering exactly this special case. In the section that follows, we shall deal with a general A_n and B_n. In the final section, we shall consider certain numerical approximation procedures meant for obtaining regularized approximate solutions for integral equations of the first kind.

5.2 Using an Approximation (T_n) of T

One of the well-known numerical procedures to obtain approximations for the operator equations of the second kind is the so-called **Ritz method**. In the present context, the second kind equation is (5.2), and in Ritz method for this, one looks for an element $x_{\alpha,n}^\delta$ in a finite dimensional subspace X_n of X such that

$$\langle T^* T x_{\alpha,n}^\delta + \alpha x_{\alpha,n}^\delta, \, u_n \rangle = \langle T^* y^\delta, u_n \rangle \quad \forall u_n \in X_n. \tag{5.4}$$

If $P_n : X \to X$ is the orthogonal projection onto the space X_n, then the above requirement is equivalent to that of finding $x_{\alpha,n}^\delta \in X_n$ such that

$$T_n^* T_n x_{\alpha,n}^\delta + \alpha x_{\alpha,n}^\delta = T_n^* y^\delta, \tag{5.5}$$

where

$$T_n = T P_n.$$

Thus, the Ritz method for obtaining approximate solution for (5.2) boils down to a specific case of the situation in which we use an approximation $T_n := T P_n$ of T. Note that, in the above discussion, P_n is not necessary to be of finite rank.

Exercise 5.2. Show that $x_{\alpha,n}^\delta$ is a solution of (5.4) if and only if it is a solution of (5.5), and the solution is unique.

5.2.1 *Convergence and general error estimates*

Let (T_n) be a sequence of operators in $\mathcal{B}(X, Y)$. Let

$$R_\alpha^{(n)} := (T_n^* T_n + \alpha I)^{-1} T_n^*, \quad \alpha > 0, \, n \in \mathbb{N}.$$

Then from Corollary 4.5 with T_n in place of T, we have

$$\|R_\alpha^{(n)}\| \le \frac{1}{2\sqrt{\alpha}}.$$

Thus, from the inequality (5.1), we have

$$\|\hat{x} - x_{\alpha,n}^\delta\| \le \|\hat{x} - x_\alpha\| + \frac{\delta}{2\sqrt{\alpha}} + \|(R_\alpha - R_\alpha^{(n)})y\|. \tag{5.6}$$

The following proposition shows that $\|(R_\alpha - R_\alpha^{(n)})y\|$ can be made small for small enough α and large enough n provided (T_n) is an approximation of T in certain sense. First we observe that

$$R_\alpha - R_\alpha^{(n)} = R_{\alpha,1}^{(n)} + R_{\alpha,2}^{(n)} + R_{\alpha,3}^{(n)}, \tag{5.7}$$

where

$$R_{\alpha,1}^{(n)} = (T^* - T_n^*)(TT^* + \alpha I)^{-1}, \tag{5.8}$$

$$R_{\alpha,2}^{(n)} = R_\alpha^{(n)} T_n (T_n^* - T^*)(TT^* + \alpha I)^{-1}, \tag{5.9}$$

$$R_{\alpha,3}^{(n)} = R_\alpha^{(n)} (T_n - T) R_\alpha. \tag{5.10}$$

In order to see the relation (5.7), first we note that

$$\begin{aligned} R_\alpha - R_\alpha^{(n)} &= (T^*T + \alpha I)^{-1} T^* - (T_n^* T_n + \alpha I)^{-1} T_n^* \\ &= T^*(TT^* + \alpha I)^{-1} - T_n^*(T_n T_n^* + \alpha I)^{-1} \\ &= (T^* - T_n^*)(TT^* + \alpha I)^{-1} \\ &\quad + T_n^*[(TT^* + \alpha I)^{-1} - (T_n T_n^* + \alpha I)^{-1}]. \end{aligned}$$

Now, (5.7) follows by writing $(TT^* + \alpha I)^{-1} - (T_n T_n^* + \alpha I)^{-1}$ as

$$(T_n T_n^* + \alpha I)^{-1}(T_n T_n^* - TT^*)(TT^* + \alpha I)^{-1}$$

and observing

$$T_n T_n^* - TT^* = T_n(T_n^* - T^*) + (T_n - T)T^*.$$

Throughout this chapter we assume that

$$y \in D(T^\dagger) \quad \text{and} \quad \|y - y^\delta\| \le \delta.$$

Proposition 5.1. *For every $\alpha > 0$ and $n \in \mathbb{N}$,*

$$\|(R_\alpha - R_\alpha^{(n)})y\| \le 2\|T - T_n\|\|(TT^* + \alpha I)^{-1}y\| + \frac{\|(T - T_n)x_\alpha\|}{2\sqrt{\alpha}}$$

$$\le \frac{\|(T - T_n)\hat{x}\|}{2\sqrt{\alpha}} + \frac{\|T - T_n\|}{\sqrt{\alpha}}\eta_\alpha(y),$$

where

$$\eta_\alpha(y) := 2\|\sqrt{\alpha}(TT^* + \alpha I)^{-1}y\| + \frac{1}{2}\|\hat{x} - x_\alpha\|.$$

In particular, if $y \in R(T)$, then $\eta_\alpha(y) \to 0$ as $\alpha \to 0$.

Proof. Let $R_{\alpha,1}^{(n)}$, $R_{\alpha,2}^{(n)}$ and $R_{\alpha,3}^{(n)}$ be as in (5.8), (5.9) and (5.10), respectively. Now, using the relations

$$\|R_\alpha^{(n)} T_n\| \leq 1, \quad \|R_\alpha^{(n)}\| \leq \frac{1}{2\sqrt{\alpha}}, \quad \|T^* - T_n^*\| = \|T - T_n\|$$

we obtain

$$\|R_{\alpha,1}^{(n)} y\| \leq \|T - T_n\| \|(TT^* + \alpha I)^{-1} y\|,$$

$$\|R_{\alpha,2}^{(n)} y\| \leq \|T - T_n\| \|(TT^* + \alpha I)^{-1} y\|,$$

$$\|R_{\alpha,3}^{(n)} y\| \leq \frac{\|(T_n - T) x_\alpha\|}{2\sqrt{\alpha}}$$

$$\leq \frac{\|(T_n - T)\hat{x}\|}{2\sqrt{\alpha}} + \frac{\|T_n - T\|}{2\sqrt{\alpha}} \|\hat{x} - x_\alpha\|.$$

Thus, from (5.7),

$$\|(R_\alpha - R_\alpha^{(n)}) y\| \leq \|R_{\alpha,1}^{(n)} y\| + \|R_{\alpha,2}^{(n)} y\| + \|R_{\alpha,3}^{(n)} y\|$$

$$\leq 2\|T - T_n\| \|(TT^* + \alpha I)^{-1} y\| + \frac{\|(T - T_n) x_\alpha\|}{2\sqrt{\alpha}}$$

$$\leq \frac{\|(T - T_n)\hat{x}\|}{2\sqrt{\alpha}} + \frac{\|T - T_n\|}{\sqrt{\alpha}} \eta_\alpha(y),$$

where

$$\eta_\alpha(y) := 2\|\sqrt{\alpha}(TT^* + \alpha I)^{-1} y\| + \frac{1}{2}\|\hat{x} - x_\alpha\|.$$

Next, suppose $y \in R(T)$. Then we have

$$\|\sqrt{\alpha}(TT^* + \alpha I)^{-1} y\| = \|\sqrt{\alpha}(TT^* + \alpha I)^{-1} T\hat{x}\|.$$

Let $A_\alpha := \sqrt{\alpha}(TT^* + \alpha I)^{-1} T$ for $\alpha > 0$. Recall from Lemma 4.1 and Corollary 4.4 that

$$\|(TT^* + \alpha I)^{-1} T^* T\| \leq 1 \quad \text{and} \quad \|(TT^* + \alpha I)^{-1} T\| \leq \frac{1}{2\sqrt{\alpha}}$$

for all $\alpha > 0$. Hence, $\|A_\alpha\| \leq 1/2$ for all $\alpha > 0$. Also, for $u \in R(T^*)$, if $u = T^* v$ with $v \in Y$, then

$$\|A_\alpha u\| = \|\sqrt{\alpha}(TT^* + \alpha I)^{-1} TT^* v\| \leq \sqrt{\alpha} \|v\|$$

so that $\|A_\alpha u\| \to 0$ as $\alpha \to 0$ for every $u \in R(T^*)$. Recall that $R(T^*)$ is dense in $N(T)^\perp$ and $\hat{x} \in N(T)^\perp$. Thus, by Theorem 2.8, we have

$$\|\sqrt{\alpha}(TT^* + \alpha I)^{-1} y\| = \|A_\alpha \hat{x}\| \to 0 \quad \text{as} \quad \alpha \to 0.$$

We already know that $\|\hat{x} - x_\alpha\| \to 0$ as $\alpha \to 0$. Hence, $\eta_\alpha(y) \to 0$ as $\alpha \to 0$ whenever $y \in R(T)$. $\qquad\square$

Corollary 5.1. *Suppose $y \in R(T)$. Then for every $\alpha > 0$ and $n \in \mathbb{N}$,*

$$\|(R_\alpha - R_\alpha^{(n)})y\| \leq \frac{3}{2\sqrt{\alpha}}\|\hat{x}\|\,\|T - T_n\|.$$

Proof.   By Proposition 5.1 we have

$$\|(R_\alpha - R_\alpha^{(n)})y\| \leq 2\|T - T_n\|\|(TT^* + \alpha I)^{-1}y\| + \frac{\|(T - T_n)x_\alpha\|}{2\sqrt{\alpha}}$$

for all $\alpha > 0$ and $n \in \mathbb{N}$. Now, if $y \in R(T)$ then $y = T\hat{x}$ so that

$$\|(TT^* + \alpha I)^{-1}y\| = \|(TT^* + \alpha I)^{-1}T\hat{x}\| \leq \frac{\|\hat{x}\|}{2\sqrt{\alpha}}$$

and

$$\|(T - T_n)x_\alpha\| = \|(T - T_n)(T^*T + \alpha I)^{-1}T^*T\hat{x}\| \leq \|\hat{x}\|.$$

From these, the required result follows. $\square$

Results in the following two theorems can be derived by applying Proposition 5.1 and Corollary 5.1 to the inequality (5.6).

Theorem 5.1. *Let $y \in R(T)$ and (T_n) be a sequence in $\mathcal{B}(X,Y)$. For $\delta > 0$, let $\alpha_\delta > 0$ be such that $\alpha_\delta \to 0$ and $\delta/\sqrt{\alpha_\delta} \to 0$ as $\delta \to 0$. Further let $n_\delta \in \mathbb{N}$ be such that*

$$\|(T - T_{n_\delta})\hat{x}\| = o(\sqrt{\alpha_\delta}),$$
$$\|T - T_{n_\delta}\| = O(\sqrt{\alpha_\delta})$$

as $\delta \to 0$. Then

$$\|\hat{x} - x_{\alpha_\delta, n_\delta}^\delta\| \to 0 \quad as \quad \delta \to 0.$$

Theorem 5.2. *Let $y \in R(T)$ and (T_n) be a sequence in $\mathcal{B}(X,Y)$. Let (ε_n) be a sequence of positive real numbers such that*

$$\|T - T_n\| \leq \varepsilon_n \quad \forall\, n \in \mathbb{N}.$$

Then

$$\|\hat{x} - x_{\alpha, n}^\delta\| \leq \|\hat{x} - x_\alpha\| + c\left(\frac{\delta + \varepsilon_n}{\sqrt{\alpha}}\right),$$

where $c = \frac{3}{2}\max\{1, \|\hat{x}\|\}$.

Remark 5.1. We may observe that the conditions that one has to impose on α and n so as to obtain convergence from the estimate in Theorem 5.2 are stronger than the conditions in Theorem 5.1. $\diamond$

Exercise 5.3. Write detailed proofs of Theorems 5.1 and 5.2.

Exercise 5.4. Prove that

$$\|(R_\alpha - R_\alpha^{(n)})y\| \leq \frac{\|(T - T_n)\hat{x}\|}{2\sqrt{\alpha}} + \|\hat{x} - x_\alpha\|\eta_{\alpha,n}$$

where

$$\eta_{\alpha,n} := \frac{\|(T^* - T_n^*)T\|}{\alpha} + \frac{\|T - T_n\|}{\sqrt{\alpha}}.$$

Hint: Use (5.7).

5.2.2 *Error estimates under source conditions*

Suppose φ is a source function for T. As in Section 4.6.2, consider the source set

$$M_{\varphi,\rho} := \{[\varphi(T^*T)]^{1/2}u : \|u\| \leq \rho\}$$

for some $\rho > 0$. Assume further that

$$\sup_{0 < \lambda \leq a} \frac{\lambda\sqrt{\varphi(\lambda)}}{\lambda + \alpha} \leq \sqrt{\varphi(\alpha)}, \qquad (5.11)$$

where $a \geq \|T\|^2$. Then, by Theorem 4.31, we have

$$\|\hat{x} - x_\alpha\| \leq \rho\sqrt{\varphi(\alpha)} \qquad (5.12)$$

whenever $\hat{x} \in M_{\varphi,\rho}$. Recall from Lemma 4.8 that if φ is concave, then it satisfies (5.11).

Now, from Theorem 5.2 we shall deduce an estimate for $\|\hat{x} - \tilde{x}_{\alpha,n}\|$ by choosing the parameter α appropriately.

Theorem 5.3. *Suppose $\hat{x} \in M_{\varphi,\rho}$. Let (ε_n) be a sequence of positive real numbers such that*

$$\|T - T_n\| \leq \varepsilon_n \quad \forall n \in \mathbb{N}.$$

Let

$$\alpha := \alpha_{\delta,n} := \varphi^{-1}[\psi^{-1}((\delta + \varepsilon_n^2)/\rho^2)].$$

Then

$$\|\hat{x} - \tilde{x}_{\alpha,n}\| \leq 2c\rho\sqrt{\psi^{-1}\left(\frac{(\delta + \varepsilon_n)^2}{\rho^2}\right)}$$

for all $n \in \mathbb{N}$, where $\psi(\lambda) = \lambda\varphi^{-1}(\lambda)$ and c is as in Theorem 5.2.

Proof. By Theorem 5.2 and the inequality (5.12), we have

$$\|\hat{x} - x_{\alpha,n}^{\delta}\| \leq c\left(\rho\sqrt{\varphi(\alpha)} + \frac{\delta + \varepsilon_n}{\sqrt{\alpha}}\right).$$

Now, since $\alpha := \alpha(\delta, n)$ satisfies the relation $\rho^2 \alpha \varphi(\alpha) = (\delta + \varepsilon_n)^2$, as in Lemma 4.9, we have

$$\rho^2 \varphi(\alpha) = \frac{(\delta + \varepsilon_n)^2}{\alpha^2} = \rho^2 \psi^{-1}\left(\frac{(\delta + \varepsilon_n)^2}{\rho^2}\right).$$

Thus, the result follows. □

Corollary 5.2. *Suppose $\hat{x} \in R((T^*T)^\nu)$ for some $\nu \in (0, 1]$ and*

$$\alpha(\delta, n) = (\delta + \varepsilon_n)^{\frac{2}{2\nu+1}}.$$

Then

$$\|\hat{x} - x_{\alpha,n}^{\delta}\| \leq c_0 (\delta + \varepsilon_n)^{\frac{2\nu}{2\nu+1}}$$

for some $c_0 > 0$.

Proof. Follows from Theorem 5.3 by taking $\varphi(\lambda) = \lambda^{2\nu}$. □

Remark 5.2. In Theorem 5.3, if we choose $\alpha := \alpha(\delta, n)$ satisfying

$$\kappa_1 (\delta + \varepsilon_n)^2 \leq \rho^2 \alpha \varphi(\alpha) \leq \kappa_2 (\delta + \varepsilon_n)^2$$

for some $\kappa_1, \kappa_2 > 0$, then we get the estimate as

$$\|\hat{x} - x_{\alpha,n}^{\delta}\| \leq 2c\rho\sqrt{\frac{1}{\kappa_1}\,\psi^{-1}\left(\frac{\kappa_2(\delta + \varepsilon_n)^2}{\rho^2}\right)},$$

where c is as in Theorem 5.2. ◇

Exercise 5.5. Prove the assertion in Remark 5.2.

5.2.3 *Error estimates for Ritz method*

We have already mentioned that if (P_n) in $\mathcal{B}(X)$ is a sequence of orthogonal projection operators on X, then the choice $T_n = TP_n$ leads to the Ritz method for equation (5.2). In this case, we have

$$T - T_n = T(I - P_n).$$

Thus, if T is a compact operator and if $P_n x \to x$ for every $x \in X$, then we have

$$\|T - T_n\| = \|(I - P_n)T^*\| \to 0.$$

It can be seen that if the sequence (X_n) of subspaces has the property that

$$X_n \subseteq X_{n+1}, \quad n = 1, 2, \ldots,$$

and

$$\cup_{n=1}^{\infty} X_n \text{ dense in } X,$$

and if P_n is the orthogonal projection onto X_n, then we do have

$$P_n x \to x \quad \text{as} \quad n \to \infty$$

for every $x \in X$.

Exercise 5.6. Prove the assertions in the above sentence.

In the setting of the above special choice of the operators T_n, we deduce an error bound which is different from the one we obtained in the last section. Throughout this subsection we assume that (ε_n) is a sequence of positive real numbers such that

$$\|T(I - P_n)\| \leq \varepsilon_n \quad \forall n \in \mathbb{N}.$$

Theorem 5.4. *Suppose* $y \in R(T)$. *Then*

$$\|\hat{x} - x_{\alpha,n}^{\delta}\| \leq \left(3 + \frac{\varepsilon_n}{2\sqrt{\alpha}}\right) (\|\hat{x} - x_\alpha\| + \|(I - P_n)\hat{x}\|) + \frac{\delta}{2\sqrt{\alpha}}.$$

Proof. From inequality (5.6), we have

$$\|\hat{x} - \tilde{x}_{\alpha,n}\| \leq \|\hat{x} - \tilde{x}_\alpha\| + \frac{\delta}{2\sqrt{\alpha}} + \|(R_\alpha - R_{\alpha,n})y\|.$$

The required inequality would follow once we prove

$$\|(R_\alpha - R_{\alpha,n})y\| \leq \left(2 + \frac{\varepsilon_n}{2\sqrt{\alpha}}\right) (\|\hat{x} - x_\alpha\| + \|(I - P_n)\hat{x}\|).$$

Let $R_{\alpha,1}^{(n)}$, $R_{\alpha,2}^{(n)}$ and $R_{\alpha,3}^{(n)}$ be as in (5.8), (5.9) and (5.10), respectively. Since $T_n = TP_n$, we have

$$R_{\alpha,1}^{(n)}y = (I - P_n)T^*(TT^* + \alpha I)^{-1}y = (I - P_n)x_a,$$

$$R_{\alpha,2}^{(n)}y = R_{\alpha,n}T_n(P_n - I)T^*(TT^* + \alpha I)^{-1}y$$

$$= R_{\alpha,n}T_n(P_n - I)x_\alpha,$$

$$R_{\alpha,3}^{(n)}y = R_{\alpha,n}(T_n - T)x_\alpha = R_{\alpha,n}T(P_n - I)x_\alpha.$$

Thus, using the relations

$$\|R_{\alpha,n}\| \leq \frac{1}{2\sqrt{\alpha}},$$

$$\|(I - P_n)x_\alpha\| \leq \|(I - P_n)\hat{x}\| + \|\hat{x} - x_\alpha\|,$$

$$\|T(P_n - I)x_\alpha\| \leq \|T(P_n - I)\| (\|(I - P_n)\hat{x}\| + \|\hat{x} - x_\alpha\|),$$

we obtain

$$\|R_{\alpha,1}^{(n)}y\| \leq \|(I - P_n)\hat{x}\| + \|\hat{x} - x_\alpha\|,$$

$$\|R_{\alpha,1}^{(n)}y\| \leq \|(I - P_n)\hat{x}\| + \|\hat{x} - x_\alpha\|,$$

$$\|R_{\alpha,1}^{(n)}y\| = \|R_{\alpha,n}T(P_n - I)x_\alpha\|$$

$$\leq \frac{\|T(I - P_n)\|}{2\sqrt{\alpha}} \left(\|(I - P_n)\hat{x}\| + \|\hat{x} - x_\alpha\|\right).$$

Thus, we obtain

$$\|(R_\alpha - R_{\alpha,n})y\| \leq \left(2 + \frac{\varepsilon_n}{2\sqrt{\alpha}}\right) \left(\|\hat{x} - x_\alpha\| + \|(I - P_n)\hat{x}\|\right).$$

This completes the proof. □

The following two corollaries are immediate from Theorem 5.4 and the inequality (5.12).

Corollary 5.3. *Suppose* $\|(I - P_n)\hat{x}\| \to 0$ *and* $\|T(I - P_n)\| \to 0$ *as* $n \to \infty$ *and suppose* (ε_n) *is such that* $\|T(I - P_n)\| \leq \varepsilon_n$ *and* $\varepsilon_n \to 0$ *as* $n \to \infty$. *If* $\alpha = \alpha(\delta, n)$ *is chosen such that*

$$\frac{\delta}{\sqrt{\alpha}} = o(1) \quad and \quad \frac{\varepsilon_n}{\sqrt{\alpha}} = O(1)$$

as $\delta \to 0$ *and* $n \to \infty$, *then*

$$\|\hat{x} - \tilde{x}_{\alpha,n}\| \to 0 \quad as \quad \delta \to 0, \; n \to \infty.$$

Corollary 5.4. *Suppose* $\hat{x} \in M_{\varphi,\rho}$, *where* φ *is as in Section 5.2.2, and suppose* $\alpha_\delta := \varphi^{-1}[\psi^{-1}(\delta^2/\rho^2)]$ *and* $n_\delta \in \mathbb{N}$ *is such that*

$$\varepsilon_n \leq c_1\sqrt{\alpha_\delta} \quad \forall n \geq n_\delta$$

for some $c_1 > 0$. *Then there exist* $c_2, c_3 > 0$ *such that*

$$\|\hat{x} - \tilde{x}_{\alpha_\delta,n}\| \leq c_2\|(I - P_n)\hat{x}\| + c_3\rho\sqrt{\psi^{-1}(\delta^2/\rho^2)}.$$

Exercise 5.7. Write detailed proofs of Corollaries 5.3 and 5.4.

Remark 5.3. Regularized approximation discussed in the above two subsections has been the subject matter of many works in the literature (see, e.g., [16, 21, 62, 63]). ◇

5.3   Using an Approximation (A_n) of T^*T

In this section we consider a more general situation in which we approximate the operators T^*T and T^* by sequences of operators (A_n) and (B_n) respectively, and consider the approximate equation (5.3). We shall derive the error estimates under the following general assumption.

Assumption 5.1. For each $\alpha > 0$, there exists a positive integer N_α such that for every $n \geq N_\alpha$, the operator $A_n + \alpha I$ is bijective. Let $c_{\alpha,n} > 0$ be an upper bound for $\|(A_n + \alpha I)^{-1}\|$, that is,

$$\|(A_n + \alpha I)^{-1}\| \leq c_{\alpha,n} \quad \forall n \geq N_\alpha. \tag{5.13}$$

In the following we use the notation $A := T^*T$.

We observe the following specific cases wherein the assumption made above is satisfied:

(i) Suppose A_n is a positive operator on X for each $n \in \mathbb{N}$. Then $A_n + \alpha I$ is bijective for every $\alpha > 0$ and $n \in \mathbb{N}$, and

$$\|(A_n + \alpha I)^{-1}\| \leq \frac{1}{\alpha} \quad \forall \alpha > 0,\ \forall n \geq N_\alpha.$$

(ii) Suppose $\|A - A_n\| \to 0$ as $n \to \infty$. For $\alpha > 0$, let $N_\alpha \in \mathbb{N}$ be such that $\|A - A_n\| \leq \alpha/2$ for all $n \geq N_\alpha$. Then, for every $n \geq N_\alpha$, $A_n + \alpha I$ is bijective and

$$\|(A_n + \alpha I)^{-1}\| \leq \frac{2}{\alpha} \quad \forall n \geq N_\alpha.$$

Exercise 5.8. Prove the assertions (i) and (ii) above.

Recall that if (T_n) is a sequence of operators in $\mathcal{B}(X, Y)$ such that $\|T - T_n\| \to 0$ as $n \to \infty$, then with $A_n := T_n^* T_n$, we have

$$\|A - A_n\| \to 0 \quad \text{as} \quad n \to \infty.$$

Moreover, for each $n \in \mathbb{N}$, A_n is a positive operator. A particular case of this situation, as we have already noted in Section 5.2.3, is $T_n := TP_n$, where T is a compact operator and (P_n) is a sequence of projections on X which converges pointwise to the identity operator. In the next subsection we shall describe another example of practical importance where (A_n) is a sequence of positive operators such that $\|A - A_n\| \to 0$ as $n \to \infty$.

Now, let us derive an estimate for the error $\|\hat{x} - x_{\alpha,n}^\delta\|$ under the Assumption 5.1. From equations (5.2) and (5.3), we obtain

$$(A_n + \alpha I)(x_\alpha^\delta - x_{\alpha,n}^\delta) = (T^* - B_n)y^\delta - (A - A_n)x_\alpha^\delta.$$

Hence, the following theorem is immediate.

Theorem 5.5. *Let Assumption 5.1 be satisfied. Then*

$$\|\hat{x} - x_{\alpha,n}^\delta\| \le \|\hat{x} - x_\alpha^\delta\| + c_{\alpha,n} \left(\|(T^* - B_n)y^\delta\| + \|(A - A_n)x_\alpha^\delta\| \right).$$

The following corollary is a consequence of the above theorem, as

$$\|\hat{x} - x_\alpha^\delta\| \le \|\hat{x} - x_\alpha\| + \delta/2\sqrt{\alpha} \quad \text{and} \quad \|y - y^\delta\| \le \delta.$$

Corollary 5.5. *Let Assumption 5.1 be satisfied. Then*

$$\|\hat{x} - x_{\alpha,n}^\delta\| \le a_n(\alpha)\|\hat{x} - x_\alpha\| + b_n(\alpha)\frac{\delta}{2\sqrt{\alpha}} + \varepsilon_n(\alpha),$$

where

$$a_n(\alpha) := 1 + c_{\alpha,n}\|A - A_n\|,$$
$$b_n(\alpha) := 1 + c_{\alpha,n}\left(2\sqrt{\alpha}\|T^* - B_n\| + \|A - A_n\|\right),$$
$$\varepsilon_n(\alpha) := c_{\alpha,n}\left(\|(T^* - B_n)y\| + \|(A - A_n)\hat{x}\|\right).$$

Remark 5.4. In view of the observations following Assumption 5.1, in Theorem 5.5 and Corollary 5.5 we can take $c_{\alpha,n} = 1/\alpha$ if A_n is a positive operator for every $n \in \mathbb{N}$, and $c_{\alpha,n} = 2/\alpha$ for every $n \ge n_\alpha$ for some $n_\alpha \in \mathbb{N}$ if $\|A - A_n\| \to 0$ as $n \to \infty$.   ◇

5.3.1   *Results under norm convergence*

In order to derive meaningful inference from the estimate in Corollary 5.5, it is necessary that the quantities $\|A - A_n\|$ and $\|T^* - B_n\|$ should be small. In the next two theorems, which are consequences of Theorem 5.5 and Corollary 5.5, respectively, we obtain convergence and certain error estimates under the assumption that $\|A - A_n\| \to 0$ and $\|T^* - B_n\| \to 0$ as $n \to \infty$ and under appropriate choices of $\alpha := \alpha(\delta)$ and $n := n_\delta$.

Theorem 5.6. *Suppose $\|A - A_n\| \to 0$ and $\|T^* - B_n\| \to 0$ as $n \to \infty$. Let $\alpha_\delta := \alpha(\delta)$ be such that*

$$\alpha_\delta \to 0 \quad \text{and} \quad \frac{\delta}{\sqrt{\alpha_\delta}} \to 0 \quad \text{as} \quad \delta \to 0,$$

and let $n_\delta \in \mathbb{N}$ be such that

$$\|A - A_{n_\delta}\| = O(\alpha_\delta), \quad \|T^* - B_{n_\delta}\| = O(\sqrt{\alpha_\delta}),$$
$$\|(T^* - B_{n_\delta})y\| = o(\alpha_\delta), \quad \|(A - A_{n_\delta})\hat{x}\| = o(\alpha_\delta).$$

Then

$$\|\hat{x} - x_{\alpha_\delta,n_\delta}^\delta\| \to 0 \quad \text{as} \quad \delta \to 0.$$

Theorem 5.7. *Suppose* $\|A - A_n\| \to 0$ *and* $\|T^* - B_n\| \to 0$ *as* $n \to \infty$. *Let* c, c_1, c_2 *be positive constants such that for each* $\alpha > 0$, *there exists* $N_\alpha \in \mathbb{N}$ *such that*

$$\|A - A_n\| \leq c_1\alpha, \quad \|T^* - B_n\| \leq c_2\sqrt{\alpha}, \quad \|(A_n + \alpha I)^{-1}\| \leq \frac{c}{\alpha}$$

for all $n \geq N_\alpha$. *Then*

$$\|\hat{x} - x_{\alpha,n}^\delta\| \leq a\|\hat{x} - x_\alpha\| + b\frac{\delta}{\sqrt{\alpha}} + c\frac{1}{\alpha}\left(\|(T^* - B_n)y\| + \|(A - A_n)\hat{x}\|\right),$$

where $a = 1 + cc_1$ *and* $b = 1 + cc_1 + cc_2$.

In particular, if A_n *is a positive self adjoint operator and* $B_n := T^*$ *for all* $n \in \mathbb{N}$, *then*

$$\|\hat{x} - x_{\alpha,n}^\delta\| \leq (1 + c)\left(\|\hat{x} - x_\alpha\| + \frac{\delta}{\sqrt{\alpha}}\right) + \frac{\|(A - A_n)\hat{x}\|}{\alpha}.$$

Remark 5.5. Error analysis of the method (5.3) in a more general setting has been carried out in [48] and [65]. $\lozenge$

Exercise 5.9. Write detailed proofs of Theorems 5.6 and 5.7.

Now, let us obtain some error estimates under a general source condition on $\hat{x}$.

Theorem 5.8. *Suppose* $\|A - A_n\| \to 0$ *and* $\|T^* - B_n\| \to 0$ *as* $n \to \infty$, *and* $\hat{x} \in M_{\varphi,\rho}$, *where* φ *is a source function for* T *satisfying the relation* (5.11) *as in Section 5.2.2. Suppose*

$$\alpha_\delta := \varphi^{-1}[\psi^{-1}(\delta^2/\rho^2)]$$

and $n_\delta \in \mathbb{N}$ *is such that*

$$\|A - A_{n_\delta}\| = O(\alpha_\delta), \quad \|T^* - B_{n_\delta}\| = O(\sqrt{\alpha_\delta}),$$

$$\|(T^* - B_{n_\delta})y\| + \|(A - A_{n_\delta})\hat{x}\| = O(\delta\sqrt{\alpha_\delta}).$$

Then

$$\|\hat{x} - x_{\alpha_\delta,n_\delta}^\delta\| = O\left(\sqrt{\psi^{-1}(\delta^2/\rho^2)}\right).$$

Proof. By Theorem 5.7, the assumptions on A_{n_δ} and B_{n_δ} imply that

$$\|\hat{x} - x_{\alpha_\delta,n_\delta}^\delta\| \leq \tilde{c}\left(\|\hat{x} - x_{\alpha_\delta}\| + \frac{\delta}{\sqrt{\alpha}}\right)$$

for some $\tilde{c} > 0$. Now, recall from Theorem 4.31 that $\|\hat{x} - x_{\alpha_\delta}\| \leq \rho\sqrt{\varphi(\alpha_\delta)}$. Hence, the result follow from Lemma 4.9. $\square$

Exercise 5.10. Write details of the proof of Theorem 5.8.

5.4   Methods for Integral Equations

In this section, we shall consider two methods for integral equations of the first kind,

$$\int_a^b k(s,t)x(t)\,dt = y(s), \quad s \in [a,b],$$

where $k(\cdot,\cdot)$ is a continuous function defined on $[a,b] \times [a,b]$ and $y \in L^2[a,b]$. The first one is a *degenerate kernel method*, proposed by Groetsch in [27], obtained by approximating the kernel of the integral operator $A := T^*T$ by a convergent quadrature rule, and the second one is obtained by using the Nyström approximation of A. In both methods we shall obtain error estimates in terms of the uniform norm $\|\cdot\|_\infty$ as well as the L^2-norm $\|\cdot\|_2$.

We have seen in Chapters 2 and 4 that the operator T defined by

$$(Tx)(s) = \int_a^b k(s,t)x(t)dt, \quad x \in X, \ s \in [a,b],$$

is a compact operator from X to Y, where X and Y are any of the spaces $L^2[a,b]$ with $\|\cdot\|_2$ and $C[a,b]$ with $\|\cdot\|_\infty$. The same is the case with the integral operators T^* and T^*T. Note that T^* and T^*T are defined by

$$(T^*x)(s) = \int_a^b \overline{k(t,s)}x(t)dt, \quad x \in L^2[a,b],$$

and

$$(T^*Tx)(s) = \int_a^b \tilde{k}(s,t)x(t)dt, \quad x \in L^2[a,b],$$

respectively, where

$$\tilde{k}(s,t) = \int_a^b \overline{k(\tau,s)}k(\tau,t)d\tau, \quad s,t \in [a,b].$$

As in earlier sections, we assume that for $\delta > 0$, $y^\delta \in L^2[a,b]$ is such that

$$\|y - y^\delta\|_2 \le \delta.$$

We also recall that from Theorem 4.10 and Lemma 4.3 that,

$$\|\hat{x} - x_\alpha\|_2 \to 0 \quad \text{as} \quad \alpha \to 0$$

and

$$\|x_\alpha - x_\alpha^\delta\|_2 \le \frac{\delta}{2\sqrt{\alpha}}$$

for $y \in D(T^\dagger)$, where

$$\hat{x} := T^\dagger y, \quad x_\alpha := (T^*T + \alpha I)^{-1}T^*y, \quad x_\alpha^\delta := (T^*T + \alpha I)^{-1}T^*y^\delta.$$

However, with respect to the uniform $\|\cdot\|_\infty$, we only have the following theorem (cf. Groetwsch [27]), for the proof of which we shall make use of the relation

$$\|T^*x\|_\infty \le \kappa_0\|x\|_2, \tag{5.14}$$

where

$$\kappa_0 := \left(\sup_{a \le s \le b} \int_a^b |(k(s,t)|^2 dt \right)^{1/2}.$$

Since $k(\cdot,\cdot)$ is continuous, we know that for every $x \in L^2[a,b]$, the functions Tx, T^*x and T^*Tx are in $C[a,b]$. In particular, it follows that x_α and x_α^δ are in $C[a,b]$.

Exercise 5.11. Justify the assertion in the last statement.

Theorem 5.9. *For $\alpha > 0$ and $\delta > 0$,*

$$\|x_\alpha - x_\alpha^\delta\|_\infty \le \kappa_0\frac{\delta}{\alpha}$$

with κ_0 as in (5.14). If $\hat{x} \in R(T^*)$, then

$$\|\hat{x} - x_\alpha\|_\infty \to 0 \quad as \quad \alpha \to 0.$$

Proof. Using the relation (5.14), we have

$$\begin{aligned}
\|x_\alpha - x_\alpha^\delta\|_\infty &= \|T^*(TT^* + \alpha I)^{-1}(y - y^\delta)\|_\infty \\
&\le \kappa_0\|(TT^* + \alpha I)^{-1}(y - y^\delta)\|_2 \\
&\le \frac{\kappa_0\delta}{\alpha}.
\end{aligned}$$

Next, suppose that $\hat{x} \in R(T^*)$. Then there exists $u \in N(T^*)^\perp$ such that $\hat{x} = T^*u$. Hence, by the relation (5.14) and the fact that $\hat{x} - x_\alpha = \alpha(T^*T + \alpha I)^{-1}\hat{x}$, we have

$$\begin{aligned}
\|\hat{x} - x_\alpha\|_\infty &= \alpha\|T^*(TT^* + \alpha I)^{-1}u\|_\infty \\
&\le \kappa_0\|\alpha(TT^* + \alpha I)^{-1}u\|_2.
\end{aligned}$$

Now, by Lemma 4.2, we have $\|\alpha(TT^* + \alpha I)^{-1}u\|_2 \to 0$ as $\alpha \to 0$. Thus, $\|\hat{x} - x_\alpha\|_\infty \to 0$ as $\alpha \to 0$. $\square$

Exercise 5.12. Derive the inequality in (5.14).

As we have already mentioned in the beginning of this section, the methods that we describe are based on a convergent quadrature rule, say with nodes $\tau_j^{(n)}$ in $[a, b]$ and weights $w_j^{(n)} \geq 0$ for $j \in \{1, \ldots, n\}$. Thus, we have

$$\sum_{j=1}^{n} x(\tau_j^{(n)}) w_j^{(n)} \rightarrow \int_a^b x(t)\, dt \quad \text{as} \quad n \rightarrow \infty$$

for every $x \in C[a, b]$.

In the following, for the simplicity of notation, we drop the superscripts in $t_j^{(n)}$ and $w_j^{(n)}$, and write them as t_j and w_j, respectively, for $n \in \mathbb{N}$ and $j = 1, \ldots, n$. The convergence in $C[a, b]$ is with respect to the norm $\| \cdot \|_\infty$, and for a bounded operator $B : L^2[a, b] \rightarrow L^2[a, b]$, $\|B\|$ denotes, as usual, the operator norm induced by the norm $\| \cdot \|_2$ on $L^2[a, b]$. If $B : C[a, b] \rightarrow C[a, b]$ is a bounded operator with respect to the norm $\| \cdot \|_\infty$ on $C[a, b]$, then we shall denote the operator norm of B by $\|B\|_\infty$. We shall also assume, without loss of generality, that $b - a = 1$.

5.4.1 *A degenerate kernel method*

In this method, we take A_n to be an approximation of $A := T^*T$ by approximating the kernel $\tilde{k}(\cdot, \cdot)$ of A by the quadrature formula. Thus, we define the integral operator A_n by

$$(A_n x)(s) = \int_a^b \tilde{k}_n(s, t) x(t)\, dt, \quad x \in L^2[a, b], \; s \in [a, b],$$

where

$$\tilde{k}_n(s, t) = \sum_{j=1}^{n} \overline{k(\tau_j, s)} k(\tau_j, t) w_j, \quad s, t \in [a, b].$$

Note that A_n is a *degenerate kernel approximation* of $A := T^*T$.

An important observation about the operator A_n is that if F_n is the Nyström approximation of T^* which is considered as an operator from $C[a, b]$ into itself, that is,

$$(F_n x)(s) = \sum_{j=1}^{n} \overline{k(\tau_j, s)} x(\tau_j) w_j, \quad x \in C[a, b],$$

then

$$A_n := F_n T. \tag{5.15}$$

Indeed, for $x \in L^2[a, b]$ and $s \in [a, b]$,

$$
\begin{aligned}
(A_n x)(s) &= \int_a^b \tilde{k}_n(s, t) x(t) \, dt \\
&= \int_a^b \sum_{j=1}^n \overline{k(\tau_j, s)} k(\tau_j, t) w_j x(t) \, dt \\
&= \sum_{j=1}^n \overline{k(\tau_j, s)} \left(\int_a^b k(\tau_j, t) x(t) \, dt \right) w_j \\
&= (F_n T x)(s).
\end{aligned}
$$

By Theorem 3.5, we know that for every $x \in C[a, b]$,

$$
\| T^* x - F_n x \|_\infty \to 0 \quad \text{as} \quad n \to \infty. \tag{5.16}
$$

The above observations facilitate us to prove the following result.

Proposition 5.2. *For each $n \in \mathbb{N}$, A_n is a positive operator and*

$$
\| A - A_n \| \to 0 \quad \text{as} \quad n \to \infty.
$$

Proof. We observe that, for every $x \in X$,

$$
\begin{aligned}
\langle A_n x, x \rangle &= \int_a^b (F_n T x)(s) \overline{x(s)} ds \\
&= \sum_{j=1}^n \int_a^b \overline{k(\tau_j, s)} \left(\int_a^b k(\tau_j, t) x(t) dt \right) w_j \overline{x(s)} ds \\
&= \sum_{j=1}^n w_j \int_a^b \int_a^b \overline{k(\tau_j, s)} k(\tau_j, t) x(t) \overline{x(s)} ds dt \\
&= \sum_{j=1}^n w_j \left| \int_a^b k(\tau_j, s) x(s) ds \right|^2 \\
&\geq 0.
\end{aligned}
$$

Thus, A_n is a positive operator for every $n \in \mathbb{N}$.

Since T is a compact operator from $L^2[a, b]$ to $C[a, b]$, by (5.16) and Theorem 2.13, we also have

$$
\| A - A_n \| = \| (T^* - F_n) T \| \to 0 \quad \text{as} \quad n \to \infty.
$$

This completes the proof. $\qquad\qquad\qquad\qquad\qquad\qquad\qquad\qquad\qquad\qquad$ $\square$

(i) Error estimate with respect to $\|\cdot\|_2$

By Proposition 5.2, the sequence (A_n) of operators not only converges in norm, but also satisfy the resolvent estimate

$$\|(A_n + \alpha I)^{-1}\| \leq \frac{1}{\alpha} \quad \forall \alpha > 0. \tag{5.17}$$

In particular, the equation

$$(A_n + \alpha I)x_{\alpha,n}^\delta = T^* y^\delta \tag{5.18}$$

has a unique solution for every $n \in \mathbb{N}$ and $\delta > 0$.

Thus, results in Section 5.3 can be applied. In fact, in this case, the following theorem is a consequence of Theorem 5.7.

Theorem 5.10. *For $\alpha > 0$, let $n_\alpha \in \mathbb{N}$ be such that $\|A - A_n\| \leq \alpha$ for all $n \geq n_\alpha$. Then,*

$$\|\hat{x} - x_{\alpha,n}^\delta\|_2 \leq 2\left(\|\hat{x} - x_\alpha\|_2 + \frac{\delta}{\sqrt{\alpha}}\right) + \frac{1}{\alpha}\|(A - A_n)\hat{x}\|_2.$$

In particular, if α_δ is such that $\alpha_\delta \to 0$ and $\delta/\sqrt{\alpha_\delta} \to 0$ as $\delta \to 0$, and if $n_\delta \in \mathbb{N}$ is such that $\|A - A_{n_\delta}\| = o(\alpha_\delta)$, then

$$\|\hat{x} - x_{\alpha_\delta,n_\delta}^\delta\|_2 \to 0 \quad as \quad \delta \to 0.$$

Exercise 5.13. Write details of the proof of Theorem 5.10.

As a special case of Theorem 5.8, we now obtain an error estimate under a general source condition.

Theorem 5.11. *Suppose $\hat{x} \in M_{\varphi,\rho}$, where φ is a source function satisfying the relation (5.11) as in Section 5.2.2. Suppose*

$$\alpha_\delta := \varphi^{-1}[\psi^{-1}(\delta^2/\rho^2)]$$

and $n_\delta \in \mathbb{N}$ is such that

$$\|A - A_{n_\delta}\| = O(\alpha_\delta) \quad and \quad \|(A - A_{n_\delta})\sqrt{\varphi(A)}\| = O(\alpha_\delta\sqrt{\varphi(\alpha_\delta)}).$$

Then

$$\|\hat{x} - x_{\alpha_\delta,n_\delta}^\delta\| = O\left(\sqrt{\psi^{-1}(\delta^2/\rho^2)}\right).$$

Proof. In view of Theorems 4.13 and 4.31, if $\hat{x} \in M_{\varphi,\rho}$ then $\alpha = O(\varphi(\alpha))$ so that

$$\|A - A_{n_\delta}\| = O(\alpha_\delta) = O(\varphi(\alpha_\delta)) = O(\sqrt{\varphi(\alpha_\delta)}).$$

Hence, the proof follows from Theorems 5.10 by recalling Lemma 4.9. $\square$

Remark 5.6. The degenerate kernel method (5.18) has been considered by Groetsch in [27]. However, our observation (5.15) paved the way for a simpler presentation of the analysis than the one adopted by Groetsch. Moreover, the conditions required in Theorem 5.10 are weaker than the corresponding conditions in [27]. In fact, in [27], Groetsch requires the condition $\|A - A_n\| = O(\alpha^{3/2})$ for convergence and the condition $\|A - A_n\| = O(\alpha^2)$ for error estimate (under Hölder-type source condition), whereas, we obtain convergence as well as order optimal estimate under the assumption $\|A - A_n\| = O(\alpha)$. $\diamond$

(ii) Error estimate with respect to $\|\cdot\|_\infty$

Since $R(T^*)$ and $R(A_n)$ are subsets of $C[a, b]$, we have

$$x^\delta_{\alpha,n} = \frac{1}{\alpha}(T^*y^\delta - A_n x^\delta_{\alpha,n}) \in C[a, b].$$

Recall from Theorem 5.9 that

$$\|x_\alpha - x^\delta_\alpha\|_\infty \le \kappa_0 \frac{\delta}{\alpha}$$

and if $\hat{x} \in R(T^*)$, then

$$\|\hat{x} - x_\alpha\|_\infty \to 0 \quad \text{as} \quad \alpha \to 0.$$

In order to derive estimate for the error $\|x^\delta_\alpha - x^\delta_{\alpha,n}\|_\infty$, we require an estimate for $\|(A_n + \alpha I)^{-1}\|_\infty$. We shall also make use of the convergence

$$\|A - A_n\|_\infty = \|(F_n - T^*)T\|_\infty \to 0 \quad \text{as} \quad n \to \infty$$

which follows from (5.16) by the compactness of the operator T, by using Theorem 2.13.

Exercise 5.14. Show that $\|A_n x\|_\infty \le \tilde{k}\|x\|_2$ for every $n \in \mathbb{N}$ and $x \in L^2[a, b]$, where $\tilde{\kappa} := \left(\sup_{a \le s \le b} \int_a^b |\tilde{k}(s,t)|^2 dt\right)^{1/2}$ and

$$\|(A_n + \alpha I)^{-1}\|_\infty \le \frac{1}{\alpha}\left(1 + \frac{\tilde{\kappa}}{\alpha}\right)$$

for every $\alpha > 0$ and $n \in \mathbb{N}$.
Hint: Use the identity: $(A_n + \alpha I)^{-1} = \frac{1}{\alpha}\left[I - A_n(A_n + \alpha I)^{-1}\right]$.

Now, we derive an estimate for $\|(A_n + \alpha I)^{-1}\|_\infty$ which is better than the one given in Exercise 5.14 for small α. For this purpose, first we obtain another representation for the operator A_n.

Since the quadrature formula under consideration is convergent, as a consequence of uniform boundedness principle, we know (cf. [51], Section 6.2.2) that there exists $\omega > 0$ such that

$$\sum_{j=1}^{n} w_j \leq \omega \quad \forall n \in \mathbb{N}. \tag{5.19}$$

Let $\mathbb{K}_w^n$ be the space $\mathbb{K}^n$ with the inner product

$$\langle \mathbf{a}, \mathbf{b} \rangle_w := \sum_{j=1}^{n} a_i \overline{b}_j w_j, \quad \mathbf{a}, \mathbf{b} \in \mathbb{K}^n,$$

and the corresponding induced norm $\| \cdot \|_w$, where a_j denotes the jth coordinate of $\mathbf{a}$.

Proposition 5.3. *Let* $K_n : L^2[a,b] \to \mathbb{K}_w^n$ *be defined by*

$$(K_n x)(j) := (Tx)(\tau_j), \quad x \in L^2[a,b], \quad j = 1, \dots, n.$$

Then

$$A_n = K_n^* K_n \quad \forall n \in \mathbb{N}.$$

Further, $R(K_n^*) \subseteq C[a,b]$ *and*

$$\|K_n^* \mathbf{a}\|_\infty \leq \|k\|_\infty \sqrt{\omega} \|\mathbf{a}\|_w \quad \forall a \in \mathbb{K}_w^n, \tag{5.20}$$

where $\|k\|_\infty := \sup_{s,t \in [a,b]} |k(s,t)|$.

Proof. We note that, for $x \in L^2[a,b]$ and $\mathbf{a} \in \mathbb{K}_w^n$,

$$\langle K_n x, \mathbf{a} \rangle_w = \sum_{j=1}^{n} (Tx)(\tau_j) \overline{a}_j w_j$$

$$= \int_a^b x(t) \Big(\sum_{j=1}^{n} k(\tau_j, t) \overline{a}_j w_j \Big) dt.$$

Thus, $K_n^* : \mathbb{K}_w^n \to L^2[a,b]$, the adjoint of K_n is given by

$$(K_n^* \mathbf{a})(s) := \sum_{j=1}^{n} \overline{k(\tau_j, s)} a_j w_j, \quad \mathbf{a} \in \mathbb{K}_w^n, \ s \in [a,b].$$

Clearly, $R(K_n^*) \subseteq C[a,b]$. For $x \in L^2[a,b]$ and $s \in [a,b]$, we have

$$(K_n^* K_n x)(s) = \sum_{j=1}^{n} \overline{k(\tau_j, s)} (Tx)(\tau_j) w_j$$

$$= \int_a^b x(t) \Big(\sum_{j=1}^{n} \overline{k(\tau_j, s)} k(\tau_j, t) w_j \Big) dt$$

$$= (A_n x)(s).$$

Next, we observe that for $\mathbf{a} \in \mathbb{K}_w^n$ and $s \in [a, b]$,

$$|(K_n^* a)(s)| \leq \sum_{j=1}^{n} |\overline{k(\tau_j, s)}| \, |a_j| w_j$$

$$\leq \|k\|_\infty \sum_{j=1}^{n} (|a_j| \sqrt{w_j}) \sqrt{w_j}$$

$$\leq \|k\|_\infty \Big(\sum_{j=1}^{n} |a_j|^2 w_j \Big)^{1/2} \Big(\sum_{j=1}^{n} w_j \Big)^{1/2}$$

$$\leq \|k\|_\infty \|\mathbf{a}\|_w \sqrt{\omega},$$

where ω is as in (5.19). Thus, we have proved (5.20). $\qquad \square$

Theorem 5.12. *For every $\alpha > 0$ and $n \in \mathbb{N}$,*

$$\|(A_n + \alpha I)^{-1}\|_\infty \leq \frac{1}{\alpha} \Big(1 + \frac{c_0}{2\sqrt{\alpha}} \Big), \qquad (5.21)$$

where $c_0 := \|k\|_\infty \sqrt{\omega}$ with ω is as in (5.19).

Proof. Let $x \in C[a, b]$. Using the identity

$$(A_n + \alpha I)^{-1} = \frac{1}{\alpha} \Big[I - A_n (A_n + \alpha I)^{-1} \Big],$$

the representation $A_n = K_n^* K_n$ with K_n as in Proposition 5.3, we have,

$$\|(A_n + \alpha I)^{-1} x\|_\infty \leq \frac{1}{\alpha} \big(\|x\|_\infty + \|A_n (A_n + \alpha I)^{-1} x\|_\infty \big)$$

$$\leq \frac{1}{\alpha} \big(\|x\|_\infty + \|K_n^* K_n (K_n^* K_n + \alpha I)^{-1} x\|_\infty \big).$$

Now, using the relation (5.20) in Proposition 5.3 and

$$\|K_n (K_n^* K_n + \alpha I)^{-1}\| \leq \frac{1}{2\sqrt{\alpha}},$$

we have

$$\|K_n^* K_n (K_n^* K_n + \alpha I)^{-1} x\|_\infty \leq c_0 \|K_n (K_n^* K_n + \alpha I)^{-1} x\|_w$$

$$\leq \frac{c_0}{2\sqrt{\alpha}} \|x\|_2$$

$$\leq \frac{c_0}{2\sqrt{\alpha}} \|x\|_\infty.$$

The above inequalities yield the required estimate. $\qquad \square$

Now, we are in a position to obtain an estimate for $\|\hat{x} - x_{\alpha,n}^{\delta}\|_{\infty}$.

Theorem 5.13. *For every $\alpha > 0$ and $n \in \mathbb{N}$,*

$$\|\hat{x} - x_{\alpha,n}^{\delta}\|_{\infty} \leq (1 + c_{\alpha}\|A - A_n\|_{\infty})\|\hat{x} - x_{\alpha}^{\delta}\|_{\infty} + c_{\alpha}\|(A - A_n)\hat{x}\|_{\infty},$$

where

$$c_{\alpha} := \frac{1}{\alpha}\left(1 + \frac{\|k\|_{\infty}\sqrt{\omega}}{2\sqrt{\alpha}}\right).$$

Proof.   Recall that

$$(A_n + \alpha I)(x_{\alpha}^{\delta} - x_{\alpha,n}^{\delta}) = (A_n - A)x_{\alpha}^{\delta}.$$

Hence, using the resolvent estimate (5.21) in Theorem 5.12, we have

$$\|x_{\alpha}^{\delta} - x_{\alpha,n}^{\delta}\|_{\infty} \leq c_{\alpha}\|(A_n - A)x_{\alpha}^{\delta}\|_{\infty}.$$

But,

$$\|(A_n - A)x_{\alpha}^{\delta}\|_{\infty} \leq \|A_n - A\|_{\infty}\|x_{\alpha}^{\delta} - \hat{x}\|_{\infty} + \|(A_n - A)\hat{x}\|_{\infty}.$$

The above inequalities together with the inequality

$$\|\hat{x} - x_{\alpha,n}^{\delta}\|_{\infty} \leq \|\hat{x} - x_{\alpha}^{\delta}\|_{\infty} + \|x_{\alpha}^{\delta} - x_{\alpha,n}^{\delta}\|_{\infty}$$

give the required estimate.   □

Theorem 5.13 together with Theorem 5.9 yield the following convergence result.

Corollary 5.6. *Suppose $\hat{x} \in R(T^*)$. For each $\delta > 0$, let α_{δ} be such that $\alpha_{\delta} \to 0$ and $\delta/\alpha_{\delta} \to 0$ as $\delta \to 0$ and let $n_{\delta} \in \mathbb{N}$ be such that*

$$\|A - A_{n_{\delta}}\|_{\infty} = O\left(\alpha_{\delta}^{3/2}\right), \qquad \|(A - A_{n_{\delta}})\hat{x}\|_{\infty} = o\left(\alpha_{\delta}^{3/2}\right).$$

Then

$$\|\hat{x} - x_{\alpha,n_{\delta}}^{\delta}\|_{\infty} \to 0 \quad as \quad \delta \to 0.$$

Exercise 5.15. Write details of the proof of Corollary 5.6.

Remark 5.7. In the case when B_n is not necessarily equal to T^*, we have the equation

$$(A_n + \alpha I)\tilde{x}_{\alpha,n}^{\delta} = B_n y^{\delta}. \tag{5.22}$$

For obtaining error estimate in terms of the norm $\|\cdot\|_{\infty}$, we consider operators B_n to be bounded operators from $L^2[a, b]$ to $C[a, b]$ and denote their norm by $\|B_n\|_*$. In applications, one of the choices of B_n can be of the form

$$B_n := P_n T^*,$$

where $P_n : C[a,b] \to C[a,b]$ is a bounded projection operator with respect to the norm $\|\cdot\|_\infty$ on $C[a,b]$ such that

$$\|P_n x - x\|_\infty \to 0 \quad \text{as} \quad n \to \infty.$$

In this particular case, we also have

$$\|T^* - B_n\| \to 0 \quad \text{and} \quad \|T^* - B_n\|_* \to 0 \quad \text{as} \quad n \to \infty.$$

Note that

$$\tilde{x}_{\alpha,n}^\delta = x_{\alpha,n}^\delta + z_{\alpha,n}^\delta,$$

where $x_{\alpha,n}^\delta$ is as in (5.18) and

$$z_{\alpha,n}^\delta = (A_n + \alpha I)^{-1}(B_n - T^*)y^\delta.$$

Since

$$\tilde{x}_{\alpha,n}^\delta - \hat{x} = (x_{\alpha,n}^\delta - \hat{x}) + z_{\alpha,n}^\delta,$$

it is enough to obtain estimates for $\|z_{\alpha,n}^\delta\|_2$ and $\|z_{\alpha,n}^\delta\|_\infty$. Now, using the bounds for the norms of $(A_n + \alpha I)^{-1}$ given in (5.17) and (5.21), we have

$$\|z_{\alpha,n}^\delta\|_2 \le \frac{1}{\alpha}\left(\|B_n - T^*\|\delta + \|(B_n - T^*)y\|_2\right),$$

and

$$\|z_{\alpha,n}^\delta\|_\infty \le \frac{1}{\alpha}\left(1 + \frac{c_0}{2\sqrt{\alpha}}\right)\left(\|B_n - T^*\|_*\delta + \|(B_n - T^*)y\|_\infty\right),$$

where $c_0 := \|k\|_\infty \sqrt{\omega}$ with ω is as in (5.19). $\diamond$

Remark 5.8. The representation $A_n = K_n^* K_n$ given in Proposition 5.3 has been used by Nair and Pereverzev in [54] to analyse a regularized collocation method wherein the operator B_n is taken to be K_n^* and the noisy data $\mathbf{y}^\delta$ is in $\mathbb{K}_w^n$ satisfying $\|y(\tau_j) - y_j^\delta\| \le \delta$ for $j = 1, \ldots, n$. $\diamond$

5.4.2   *Regularized Nyström method*

In this method we consider regularized approximation of $\hat{x}$ as the Nyström approximation of the solution of (5.2). Thus, in place of the integral operator $A := T^*T$, we consider its Nyström approximation $\widetilde{A}_n$ which is defined by

$$(\widetilde{A}_n x)(s) = \sum_{j=1}^{n} \tilde{k}(s, \tau_j) x(\tau_j) w_j$$

for every $x \in C[a, b]$ and $s \in [a, b]$ using a convergent quadrature formula based on the nodes $\tau_1, \ldots, \tau_n$ and weights $w_1, \ldots, w_n$. The resulting approximate equation is

$$(\widetilde{A}_n + \alpha I)x_{\alpha,n}^{\delta} = T^* y^{\delta}. \tag{5.23}$$

It is to be born in mind that the operator $\widetilde{A}_n$ above is defined on $C[a, b]$; not on the whole of $L^2[a, b]$. Thus, it is not apparent that equation (5.23) is uniquely solvable. So, our first attempt is to prove the unique solvability of (5.23). For doing this, first we observe that $\widetilde{A}_n$ can be represented as

$$\widetilde{A}_n := T^*T_n, \tag{5.24}$$

where T_n is the Nyström approximation of T, that is,

$$(T_n x)(s) = \sum_{j=1}^{n} k(s, \tau_j) x(\tau_j) w_j$$

for every $x \in C[a, b]$, $s \in [a, b]$. Indeed, for $x \in C[a, b]$ and $s \in [a, b]$,

$$
\begin{aligned}
(\widetilde{A}_n x)(s) &= \sum_{j=1}^{n} \tilde{k}(s, \tau_j) x(\tau_j) w_j \\
&= \sum_{j=1}^{n} \Big(\int_a^b \overline{k(t, s)} k(t, \tau_j) \, dt \Big) x(\tau_j) w_j \\
&= \int_a^b \overline{k(t, s)} (T_n x)(t) \, dt \\
&= (T^*T_n x)(s).
\end{aligned}
$$

As in Proposition 5.2, it can be shown that $T_n T^*$ is a positive operator on $L^2[a, b]$. In particular, $T_n T^* + \alpha I$ is bijective from $L^2[a, b]$ to itself, and hence

$$\|(T_n T^* + \alpha I)^{-1}\| \le \frac{1}{\alpha}. \tag{5.25}$$

Theorem 5.14. *For every $\alpha > 0$ and $n \in \mathbb{N}$, the operators $T_n T^* + \alpha I$ and $T^* T_n + \alpha I$ are bijective from $C[a, b]$ into itself. Further, for every $v \in C[a, b]$,*

$$(T^* T_n + \alpha I)^{-1} v = \frac{1}{\alpha} \left(v - T^* (T_n T^* + \alpha I)^{-1} T_n v \right)$$

and

$$(T_n T^* + \alpha I)^{-1} v = \frac{1}{\alpha} \left(v - T_n (T^* T_n + \alpha I)^{-1} T^* v \right).$$

Proof. Let $v \in C[a, b]$, $n \in \mathbb{N}$ and $\alpha > 0$. We have already observed that $T_n T^* + \alpha I$ is bijective from $L^2[a, b]$ to itself. Let u_n be the unique element in $L^2[a, b]$ such that

$$(T_n T^* + \alpha I) u_n = v.$$

Then,

$$u_n = \frac{1}{\alpha} (v - T_n T^* u_n) \in C[a, b].$$

Thus, $T_n T^* + \alpha I : C[a, b] \to C[a, b]$ is bijective. Hence, we also see that for $v \in C[a, b]$,

$$x_n := \frac{1}{\alpha} \left(v - T^* (T_n T^* + \alpha I)^{-1} T_n v \right)$$

is the unique element in $C[a, b]$ such that

$$(T^* T_n + \alpha I) x_n = v$$

so that $T^* T_n + \alpha I : C[a, b] \to C[a, b]$ is also bijective. Remaining part of the theorem can be verified easily. $\square$

The following corollary is obvious from the above theorem.

Corollary 5.7. *For every $\alpha > 0$ and $v \in L^2[a, b]$, there exists a unique $x_n \in C[a, b]$ such that*

$$(T^* T_n + \alpha I) x_n = T^* v.$$

Exercise 5.16. Write the proof of Corollary 5.7.

In view of Corollary 5.7, there exists a unique $x^\delta_{\alpha, n} \in C[a, b]$ such that

$$(T^* T_n + \alpha I) x^\delta_{\alpha, n} = T^* y^\delta. \tag{5.26}$$

We observe that

$$x^\delta_{\alpha, n} = T^* u^\delta_{\alpha, n},$$

where $u_{\alpha,n}^\delta \in L^2[a, b]$ is the unique element satisfying the equation

$$(T_n T^* + \alpha I) u_{\alpha,n}^\delta = y^\delta. \tag{5.27}$$

We also have

$$x_\alpha^\delta = T^* u_\alpha^\delta, \quad x_\alpha = T^* u_\alpha,$$

where

$$u_\alpha^\delta = (TT^* + \alpha I)^{-1} y^\delta, \qquad u_\alpha = (TT^* + \alpha I)^{-1} y.$$

We shall use the following notations:

$$\varepsilon_n := \|(T - T_n)T^*\|, \qquad \eta_n := \|(T - T_n)T^*T\|.$$

Let us recall from Chapter 3 that (T_n) converges pointwise to T on $C[a, b]$. Hence, by compactness of T^* and $A := T^*T$, it follows from Theorem 2.13 that

$$\varepsilon_n := \|(T - T_n)T^*\| \to 0 \quad \text{and} \quad \eta_n := \|(T - T_n)A\| \to 0$$

as $n \to \infty$.

Proposition 5.4. *For $\alpha > 0$ and $n \in \mathbb{N}$,*

$$\|u_\alpha^\delta - u_{\alpha,n}^\delta\|_2 \le \frac{\varepsilon_n \delta}{\alpha^2} + \frac{\eta_n \|\hat{x} - x_\alpha\|_2}{\alpha^2}.$$

Proof.   From the definition of u_α^δ and $u_{\alpha,n}^\delta$, we have

$$
\begin{aligned}
u_\alpha^\delta - u_{\alpha,n}^\delta &= [(TT^* + \alpha I)^{-1} - (T_n T^* + \alpha I)^{-1}] y^\delta \\
&= (T_n T^* + \alpha I)^{-1} (T_n T^* - TT^*)(TT^* + \alpha I)^{-1} y^\delta \\
&= (T_n T^* + \alpha I)^{-1} (T_n - T)T^*(TT^* + \alpha I)^{-1} y^\delta,
\end{aligned}
$$

and using the fact that $\hat{x} - x_\alpha = \alpha(T^*T + \alpha I)^{-1}\hat{x}$,

$$
\begin{aligned}
(T^*T + \alpha I)^{-1} T^* y &= (T^*T + \alpha I)^{-1} T^*T\hat{x} \\
&= T^*T(T^*T + \alpha I)^{-1}\hat{x} \\
&= \frac{1}{\alpha} T^*T(\hat{x} - x_\alpha).
\end{aligned}
$$

Hence,

$$T^*(TT^* + \alpha I)^{-1} y^\delta = T^*(TT^* + \alpha I)^{-1}(y^\delta - y) + \frac{1}{\alpha} T^*T(\hat{x} - x_\alpha).$$

Thus, using (5.25),

$$
\begin{aligned}
\|u_\alpha^\delta - u_{\alpha,n}^\delta\|_2 &= \|(T_n T^* + \alpha I)^{-1}(T_n - T)T^*(TT^* + \alpha I)^{-1}y^\delta\| \\
&\leq \frac{1}{\alpha}\|(T_n - T)T^*(TT^* + \alpha I)^{-1}y^\delta\|_2 \\
&\leq \frac{1}{\alpha}\|(T_n - T)T^*(TT^* + \alpha I)^{-1}(y^\delta - y)\|_2 \\
&\quad + \frac{1}{\alpha^2}\|(T_n - T)T^*T(\hat{x} - x_\alpha)\|_2 \\
&\leq \frac{\delta}{\alpha^2}\|(T_n - T)T^*\| + \frac{1}{\alpha^2}\|(T_n - T)A\|\,\|\hat{x} - x_\alpha\|_2.
\end{aligned}
$$

This competes the proof. □

Now, we deduce estimates for the error in the regularized approximation with respect to the norms $\|\cdot\|_2$ and $\|\cdot\|_\infty$.

(i) Error estimate with respect to $\|\cdot\|_2$

From the observations

$$
\|x_\alpha - x_\alpha^\delta\|_2 \leq \frac{\delta}{\sqrt{\alpha}} \quad \text{and} \quad \|x_\alpha^\delta - x_{\alpha,n}^\delta\|_2 \leq \|T^*\|\,\|u_\alpha^\delta - u_{\alpha,n}^\delta\|_2,
$$

we obtain the following theorem using the bound for $\|u_\alpha^\delta - u_{\alpha,n}^\delta\|_2$ obtained in Proposition 5.4.

Theorem 5.15. *Let $\kappa := \|T^*\|$. Then for every $\alpha > 0$ and $n \in \mathbb{N}$,*

$$
\|\hat{x} - x_{\alpha,n}^\delta\|_2 \leq \left(1 + \frac{\kappa \eta_n}{\alpha^2}\right)\|\hat{x} - x_\alpha\|_2 + \left(1 + \frac{\kappa \varepsilon_n}{\alpha^{3/2}}\right)\frac{\delta}{\sqrt{\alpha}}.
$$

As a consequence of Theorem 5.15, we have the following theorem.

Theorem 5.16. *Let $\kappa := \|T^*\|$. For $\alpha > 0$, let $n_\alpha \in \mathbb{N}$ be such that*

$$
\varepsilon_n := \|(T - T_n)T^*\| \leq \frac{\alpha^{3/2}}{\kappa} \quad \text{and} \quad \eta_n := \|(T - T_n)A\| \leq \frac{\alpha^2}{\kappa}
$$

for all $n \geq n_\alpha$. Then

$$
\|\hat{x} - x_{\alpha,n}^\delta\|_2 \leq 2\left(\|\hat{x} - x_\alpha\|_2 + \frac{\delta}{\sqrt{\alpha}}\right)
$$

for all $n \geq n_\alpha$. In particular, if $\alpha_\delta > 0$ is such that $\alpha_\delta \to 0$ and $\delta/\sqrt{\alpha_\delta} \to 0$ as $\delta \to 0$, and if $n_\delta := n_{\alpha_\delta}$, then

$$
\|\hat{x} - x_{\alpha,n_\delta}^\delta\|_2 \to 0 \quad \text{as} \quad \delta \to 0.
$$

Next we deduce another estimate for $\|\hat{x} - x_{\alpha,n}^{\delta}\|_2$ which leads to a result under weaker conditions on $\|(T-T_n)A\|$ and $\|(T-T_n)T^*\|$ than in Theorem 5.16.

Theorem 5.17. *For every $\alpha > 0$ and $n \in \mathbb{N}$,*

$$\|\hat{x} - x_{\alpha,n}^{\delta}\|_2 \leq \tilde{a}_n(\alpha)\|\hat{x} - x_\alpha\|_2 + \tilde{b}_n(\alpha)\frac{\delta}{\sqrt{\alpha}},$$

where

$$\tilde{a}_n(\alpha) := 1 + \frac{\eta_n}{\alpha^{3/2}}\left(1 + \frac{\varepsilon_n}{\alpha}\right) \quad and \quad \tilde{b}_n(\alpha) := 1 + \frac{\varepsilon_n}{\alpha}\left(1 + \frac{\varepsilon_n}{\alpha}\right).$$

In particular, for $\alpha > 0$, if $n_\alpha \in \mathbb{N}$ is such that

$$\varepsilon_n := \|(T - T_n)T^*\| \leq \alpha \quad and \quad \eta_n := \|(T - T_n)A\| \leq \alpha^{3/2}$$

for all $n \geq n_\alpha$, then

$$\|\hat{x} - x_{\alpha,n}^{\delta}\|_2 \leq 3\left(\|\hat{x} - x_\alpha\|_2 + \frac{\delta}{\sqrt{\alpha}}\right).$$

Proof. Using the definitions

$$u_\alpha^\delta := (TT^* + \alpha I)^{-1}y^\delta, \qquad u_{\alpha,n}^\delta := (T_nT^* + \alpha I)^{-1}y^\delta,$$

and the resolvent identity

$$(TT^* + \alpha I)^{-1} - (T_nT + \alpha I)^{-1} = (TT^* + \alpha I)^{-1}(T_n - T)T^*(T_nT^* + \alpha I)^{-1},$$

we have

$$\begin{aligned}
\|x_\alpha^\delta - x_{\alpha,n}^\delta\|_2 &= \|T^*(u_\alpha^\delta - u_{\alpha,n}^\delta\|_2 \\
&= \|T^*(TT^* + \alpha I)^{-1}(T_n - T)T^* u_{\alpha,n}^\delta\|_2 \\
&\leq \frac{1}{\sqrt{\alpha}}\|(T_n - T)T^* u_{\alpha,n}^\delta\|_2 \\
&\leq \frac{1}{\sqrt{\alpha}}\left(\varepsilon_n\|u_{\alpha,n}^\delta - u_\alpha^\delta\|_2 + \|(T_n - T)T^* u_\alpha^\delta\|_2\right).
\end{aligned}$$

Note that

$$\begin{aligned}
\|(T_n - T)T^* u_\alpha^\delta\|_2 &= \|(T_n - T)T^*(TT^* + \alpha I)^{-1}y^\delta\|_2 \\
&= \frac{\varepsilon_n\delta}{\alpha} + \|(T_n - T)T^*(TT^* + \alpha I)^{-1}y\|_2.
\end{aligned}$$

Now, since $\hat{x} - x_\alpha = \alpha(T^*T + \alpha I)^{-1}\hat{x}$, we have

$$\begin{aligned}
(T^*T + \alpha I)^{-1}T^*y &= (T^*T + \alpha I)^{-1}T^*T\hat{x} \\
&= T^*T(T^*T + \alpha I)^{-1}\hat{x} \\
&= \frac{1}{\alpha}A(\hat{x} - x_\alpha)
\end{aligned}$$

so that

$$\|(T_n - T)T^*(TT^* + \alpha I)^{-1}y\|_2 = \|(T_n - T)(T^*T + \alpha I)^{-1}T^*y)\|_2$$
$$= \frac{1}{\alpha}\|(T_n - T)A(\hat{x} - x_\alpha)\|_2$$
$$\leq \frac{\eta_n}{\alpha}\|\hat{x} - x_\alpha\|_2.$$

Thus,

$$\|(T_n - T)T^*u_\alpha^\delta\|_2 = \frac{\varepsilon_n\delta}{\alpha} + \frac{\eta_n}{\alpha}\|\hat{x} - x_\alpha\|_2$$

and hence,

$$\|x_\alpha^\delta - x_{\alpha,n}^\delta\|_2 \leq \frac{1}{\sqrt{\alpha}}\left(\varepsilon_n\|u_{\alpha,n}^\delta - u_\alpha^\delta\|_2 + \|(T_n - T)T^*u_\alpha^\delta\|_2\right)$$
$$\leq \frac{1}{\sqrt{\alpha}}\left(\varepsilon_n\|u_{\alpha,n}^\delta - u_\alpha^\delta\|_2 + \frac{\varepsilon_n\delta}{\alpha} + \frac{\eta_n}{\alpha}\|\hat{x} - x_\alpha\|_2\right).$$

Using the estimate

$$\|u_\alpha^\delta - u_{\alpha,n}^\delta\|_2 \leq \frac{\varepsilon_n\delta}{\alpha^2} + \frac{\eta_n\|\hat{x} - x_\alpha\|_2}{\alpha^2}$$

obtained in Proposition 5.4, we have

$$\|x_\alpha^\delta - x_{\alpha,n}^\delta\|_2 \leq \frac{1}{\sqrt{\alpha}}\left(\frac{\varepsilon_n^2}{\alpha^2}\delta + \frac{\varepsilon_n\eta_n}{\alpha^2}\|\hat{x} - x_\alpha\|_2 + \frac{\varepsilon_n\delta}{\alpha} + \frac{\eta_n}{\alpha}\|\hat{x} - x_\alpha\|_2\right)$$
$$= \frac{\varepsilon_n}{\alpha}\left(1 + \frac{\varepsilon_n}{\alpha}\right)\frac{\delta}{\sqrt{\alpha}} + \frac{\eta_n}{\alpha^{3/2}}\left(1 + \frac{\varepsilon_n}{\alpha}\right)\|\hat{x} - x_\alpha\|_2.$$

The above inequality together with the inequality

$$\|\hat{x} - x_{\alpha,n}^\delta\|_2 \leq \|\hat{x} - x_\alpha\|_2 + \frac{\delta}{\sqrt{\alpha}} + \|x_\alpha^\delta - x_{\alpha,n}^\delta\|_2,$$

imply the required estimates. $\qquad\square$

(ii) Error estimate with respect to $\|\cdot\|_\infty$

Now we deduce an error estimate with respect to the norm $\|\cdot\|_\infty$ on $C[a, b]$. Let us first observe that, since $R(T^*) \subseteq C[a, b]$, both x_α and x_α^δ belong to $C[a, b]$.

Theorem 5.18. *Let κ_0 be as in (5.14). Then for every $\alpha > 0$ and $n \in \mathbb{N}$,*

$$\|\hat{x} - x_{\alpha,n}^\delta\|_\infty \leq \hat{a}_n(\alpha)\|\hat{x} - x_\alpha\|_\infty + \hat{b}_n(\alpha)\frac{\delta}{\alpha}$$

where

$$\hat{a}_n(\alpha) := 1 + \frac{\kappa_0\eta_n}{\alpha^2} \quad and \quad \hat{b}_n(\alpha) := \kappa_0\left(1 + \frac{\varepsilon_n}{\alpha}\right).$$

Proof. By the relation (5.14), we have

$$\|x_\alpha^\delta - x_{\alpha,n}^\delta\|_\infty = \|T^*(u_\alpha^\delta - u_{\alpha,n}^\delta)\|_\infty$$
$$\leq \kappa_0 \|u_\alpha^\delta - u_{\alpha,n}^\delta\|_2.$$

Again, by the relation (5.14),

$$\|x_\alpha - x_\alpha^\delta\|_\infty = \|T^*(TT^* + \alpha I)^{-1}(y - y^\delta)\|_\infty$$
$$\leq \kappa_0 \|(TT^* + \alpha I)^{-1}(y - y^\delta)\|_2$$
$$\leq \frac{\kappa_0 \delta}{\alpha}.$$

Thus, using the estimate in Proposition 5.4,

$$\|\hat{x} - x_{\alpha,n}^\delta\|_\infty \leq \|\hat{x} - x_\alpha\|_\infty + \|x_\alpha - x_\alpha^\delta\|_\infty + \|x_\alpha^\delta - x_{\alpha,n}^\delta\|_\infty.$$
$$\leq \|\hat{x} - x_\alpha\|_\infty + \kappa_0 \frac{\delta}{\alpha} + \kappa_0 \|u_\alpha^\delta - u_{\alpha,n}^\delta\|_2$$
$$\leq \hat{a}_n(\alpha)\|\hat{x} - x_\alpha\|_\infty + \hat{b}_n(\alpha)\frac{\delta}{\alpha}.$$

This completes the proof.   □

Theorem 5.19. *Let κ_0 be as in (5.14). For $\alpha > 0$, let $n_\alpha \in \mathbb{N}$ be such that*

$$\|(T - T_n)A\| \leq \frac{\alpha^2}{\kappa_0} \quad \text{and} \quad \|(T - T_n)T^*\| \leq \alpha$$

for all $n \geq n_\alpha$. Then

$$\|\hat{x} - x_{\alpha,n}^\delta\|_\infty \leq 2\Big(\|\hat{x} - x_\alpha\|_\infty + \kappa_0\frac{\delta}{\alpha}\Big) \quad \forall n \geq n_\alpha.$$

In particular, if $\alpha_\delta > 0$ is such that $\alpha_\delta \to 0$ and $\delta/\alpha_\delta \to 0$ as $\delta \to 0$, and if $n_\delta := n_{\alpha_\delta}$, then

$$\|\hat{x} - x_{\alpha,n_\delta}^\delta\|_\infty \to 0 \quad as \quad \delta \to 0.$$

Proof.   Follows from Theorem 5.18 and Theorem 5.9.   □

Next, let us look at the case in which B_n is not necessarily equal to T^*. In this case, we have the equation

$$(\widetilde{A}_n + \alpha I)\tilde{x}_{\alpha,n}^\delta = B_n y^\delta. \tag{5.28}$$

As in Remark 5.7, we observe that

$$\tilde{x}_{\alpha,n}^\delta = x_{\alpha,n}^\delta + \tilde{z}_{\alpha,n}^\delta,$$

where

$$\tilde{z}_{\alpha,n}^\delta = (\widetilde{A}_n + \alpha I)^{-1}(B_n - T^*)y^\delta.$$

We consider operators B_n to be bounded operators from $L^2[a,b]$ to $C[a,b]$ and denote their norm by $\|B_n\|_*$. Thus,

$$\|\tilde{z}_{\alpha,n}^\delta\|_\infty \leq \|(\tilde{A}_n + \alpha I)^{-1}\|_\infty \|(B_n - T^*)y^\delta\|_\infty,$$

where

$$\begin{aligned}\|(B_n - T^*)y^\delta\|_\infty &\leq \|(B_n - T^*)(y^\delta - y)\|_\infty + \|(B_n - T^*)y\|_2 \\ &\leq \|B_n - T^*\|_* \delta + \|(B_n - T^*)y\|_\infty.\end{aligned}$$

Now, in order to derive an estimate for $\|(\tilde{A}_n + \alpha I)^{-1}\|_\infty$, we first recall from the definition of T_n that

$$\sup\{\|T_n v\|_\infty \leq \hat{c}_0 \|v\|_\infty, \quad v \in C[a,b], \tag{5.29}$$

where $\hat{c}_0 := \|k\|_\infty \omega$ with ω is as in (5.19).

Proposition 5.5. *For every* $v \in C[a,b]$,

$$\|(\tilde{A}_n + \alpha I)^{-1} v\|_\infty \leq \frac{1}{\alpha}\left(1 + \frac{\kappa_0 \hat{c}_0}{\alpha}\right)\|v\|_\infty,$$

where κ_0 *and* $\hat{c}_0$ *are as in (5.14) and (5.29), respectively.*

Proof. Let $v \in C[a,b]$. By Theorem 5.14 and the relations (5.14) and (5.29), we have

$$\begin{aligned}\|(T^*T_n + \alpha I)^{-1} v\|_\infty &\leq \frac{1}{\alpha}\left(\|v\|_\infty + \|T^*(T_n T^* + \alpha I)^{-1} T_n v\|_\infty\right) \\ &\leq \frac{1}{\alpha}\left(\|v\|_\infty + \kappa_0 \|(T_n T^* + \alpha I)^{-1} T_n v\|_2\right) \\ &\leq \frac{1}{\alpha}\left(\|v\|_\infty + \frac{\kappa_0}{\alpha}\|T_n v\|_2\right) \\ &\leq \frac{1}{\alpha}\left(\|v\|_\infty + \frac{\kappa_0}{\alpha}\|T_n v\|_\infty\right) \\ &\leq \frac{1}{\alpha}\left(1 + \frac{\kappa_0 \hat{c}_0}{\alpha}\right)\|v\|_\infty.\end{aligned}$$

Thus, the proof is over. $\qquad\qquad\square$

Thus, from the above lemma, we have

$$\|z_{\alpha,n}^\delta\|_\infty \leq \frac{1}{\alpha}\left(1 + \frac{\kappa_0 \hat{c}_0}{\alpha}\right)\left(\|B_n - T^*\|_* \delta + \|(B_n - T^*)y\|_\infty\right).$$

Since

$$\hat{x} - \tilde{x}_{\alpha,n}^\delta = (\hat{x} - x_{\alpha,n}^\delta) + z_{\alpha,n}^\delta,$$

we obtain an estimate for the norm of $\hat{x} - \tilde{x}_{\alpha,n}^\delta$ by using estimates for the norms of $\hat{x} - x_{\alpha,n}^\delta$ and $z_{\alpha,n}^\delta$.

PROBLEMS

(1) Suppose $y \in R(T)$ and (T_n) is a sequence of operators in $\mathcal{B}(X,Y)$ such that $\|T - T_n\| \to 0$ as $n \to \infty$. Let $x_{\alpha,n}^{\delta}$ be as in Section 5.2. Then show that there exists $n_{\delta} \in \mathbb{N}$ such that for every $n \geq n_{\delta}$,

$$\|\hat{x} - x_{\alpha,n}^{\delta}\| \leq \|\hat{x} - x_{\alpha}\| + c_1\left(\frac{\delta}{\sqrt{\alpha}}\right),$$

where $c = 3\max\{1, \|\hat{x}\|\}$.

(2) Suppose $y \in R(T)$. If α satisfies $\|Tx_{\alpha}^{\delta} - y^{\delta}\| \leq \delta$ and $x_{\alpha,n}^{\delta}$ is as in Section 5.2, then show that $\|(TT^* + \alpha I)^{-1}T\hat{x}\| \leq 3\delta/\alpha$ and

$$\|\hat{x} - x_{\alpha,n}^{\delta}\| \leq \frac{\|(T - T_n)\hat{x}\|}{2\sqrt{\alpha}} + c\left(1 + \frac{\|T - T_n\|}{\sqrt{\alpha}}\right)\left(\|\hat{x} - x_{\alpha}\| + \frac{\delta}{\sqrt{\alpha}}\right)$$

for some constant $c > 0$.

(3) Suppose $\hat{x} \in M_{\varphi,\rho}$ and $\|T - T_n\| \to 0$ as $n \to \infty$ and $x_{\alpha,n}^{\delta}$ is as in Section 5.2. Let $\alpha := \alpha(\delta)$ be chosen such that

$$\kappa_1 \delta^2 \leq \rho^2 \alpha \varphi(\alpha) \leq \kappa_2 \delta^2$$

for some $\kappa_1, \kappa_2 > 0$. Then show that there exists $n_{\delta} \in \mathbb{N}$ such that

$$\|\hat{x} - x_{\alpha,n}^{\delta}\| \leq \tilde{c}\rho\sqrt{\psi^{-1}\left(\frac{\delta^2}{\rho^2}\right)} \quad \forall n \geq n_{\delta}$$

for some $\tilde{c} > 0$, where $\psi(\lambda) = \lambda\varphi^{-1}(\lambda)$.

(4) Let $\psi(\lambda) = \lambda\varphi^{-1}(\lambda)$. Then show that

$$\alpha := \alpha_{\delta,n} := \varphi^{-1}[\psi^{-1}((\delta + \varepsilon_n^2)/\rho^2)] \iff \rho^2 \alpha \varphi(\alpha) = (\delta + \varepsilon_n)^2$$

and in that case

$$\rho^2 \varphi(\alpha) = \frac{(\delta + \varepsilon_n)^2}{\alpha^2} = \rho^2 \psi^{-1}\left(\frac{(\delta + \varepsilon_n)^2}{\rho^2}\right).$$

(5) If $\|(A - A_n)A\| \to 0$ and $\|(A - A_n)A_n\| \to 0$ as $n \to \infty$, then show that for every $\alpha > 0$ there exists a positive integer N_{α} such that Assumption 5.1 is satisfied with $c_{\alpha,n} = c/\alpha^2$ for some constant $c > 0$ (independent of α and n).

(6) Obtain a result analogous to Corollary 5.5 with assumptions on

$$\|(A - A_n)A\|, \quad \|(A - A_n)A_n\| \quad \text{and} \quad \|(A - A_n)T^*\|$$

instead of assumptions on $\|A_n - A\|$.

(7) Suppose $c_{\alpha,n}$ is as in Assumption 5.1 and $x_{\alpha,n}^{\delta}$ is th unique solution of (5.3). Then show that

$$\|x_{\alpha}^{\delta} - x_{\alpha,n}^{\delta}\| \leq c_{\alpha,n} \left[\|(T^* - B_n)y^{\delta}\| + \frac{\|(A - A_n)T^*\|}{\sqrt{\alpha}} \left(\frac{\delta}{\sqrt{\alpha}} + \|\hat{x}\| \right) \right].$$

(8) Let $\kappa := \sup_{a \leq s \leq b} \left(\int_a^b |k(s,t)|^2 dt \right)^{1/2}$. Then show that for every $\alpha > 0$, $A + \alpha I$ is bijective from $C[a,b]$ to itself, and for every $x \in C[a,b]$,

$$\|(A + \alpha I)^{-1} x\|_{\infty} \leq c_{\alpha} \|x\|_{\infty},$$

where $c_{\alpha} := (\sqrt{\alpha} + \kappa/2)/\alpha^{3/2}$.

(9) Prove that for every $\alpha > 0$ and $n \in \mathbb{N}$,

$$\|(\tilde{A}_n + \alpha I)^{-1} x\|_{\infty} \leq \frac{1}{\alpha} \left(1 + \frac{c_0 \kappa_0}{\alpha} \right) \|x\|_{\infty},$$

where κ and c_0 are positive constants as in (5.29) and (5.14) respectively.

(10) Using the resolvent estimate in Proposition 5.5, derive an estimate for $\|\hat{x} - \tilde{x}_{\alpha,n}^{\delta}\|$, where $\tilde{x}_{\alpha,n}^{\delta}$ is as in (5.28).

Bibliography

[1] M. AHUES, A class of strongly stable approximations, *J. Austral. Math. Soc.*, Series B, **28** (1987) 435–442.

[2] M. AHUES, A. LARGILLIER and B.V. LIMAYE, *Spectral Computations for Bounded Operators*, Applied Mathematics, Vol. 18, Chapman & Hall/CRC, Boca Raton, xvii + 382, 2001

[3] P.M. ANESELONE, *Collectively compact Operator Approximation Theory and Applications to integral Equations*, Prentice–Hall, Englewood Cliffs, 1971.

[4] R. ARCANGELI, Psedo–solution de l'equation $Ax = y$, *Comptes Rendus del' Academie des Sciences*, Paris, **263** (1966) 282–285.

[5] K.E. ATKINSON, *The Numerical Solution of Integral Equations of the Second Kind*, Cambridge University Press, 1997

[6] G.B. BAKER and T.M. MILLS, Lagrange Interpolation – divergence, *Mathematical Chronicle*, **17** (1988) 1–18.

[7] R. BOULDIN, Operator approximations with stable eigenvalues, *J. Austral. Math. Soc.*, Series A, **49** (1990) 250–257.

[8] J. BAUMEISTER, *Stable Solutions of Inverse Problems*, Fried. Vieweg & Sohn, Braunschweig/Wiesbaden1987.

[9] H. BRACKAGE, Über die numerische Behandlung von Integralgeichungen nach der Quadraturformelmethode, *Numerische Math.* **2** (1961) 183–193.

[10] F. CHATELIN, *Spectral Approximation of Linear Operators*, Academic Press, New York, 1983.

[11] F. DE HOOG, Review of Fredholm integral equations of the first kind, In: *Applications and Numerical Solution of Integral equations*, Eds.: R.S. Anderssen, F. de Hoog and M.A. Lucas, Sijthoff & Noordhoff, 1980, Pages: 119–134.

[12] D.R. DELLWO, Accelerated refinement with applications to integral equations, *SIAM J. Numer. Anal.*, **25** (1988) 1327–1339.

[13] S. DEMKO, Spline approximation in Banach function space, In: *Theory of Approximation with Applications*, Eds.: A.G. Law and B.N. Shahney, Academic Press (1976) 146–154.

[14] R.A. DE VORE, Degree of approximation, In: *Approximation Theory II*,

Eds.: G.G. Lorentz, C.K. Chui and L.L. Schumaker, Academic Press (1976) 117–161.

[15] H.W. ENGL, Regularization methods for the stable solution of inverse problems, *Surveys Math. Indust.*, **3** (1993) 71–143.

[16] H.W. ENGL and A. NEUBAUER, An improved version of Marti's method for solving ill-posed linear integral equations, *Math. Comp.*, **45** (1985) 405–416.

[17] H.W. ENGL, M. HANKE and A. NEUBAUER, *Regularization of Inverse Problems*, Kluwer Acad. Publ., Dordrecht, Boston, London, 1996.

[18] M. FRIEDMAN and D.R. DELLWO, Accelerated projection methods, *J. Comp. Phys.*, **45** (1982) 108–120.

[19] M. FRIEDMAN and D.R. DELLWO, Accelerated quadrature methods for linear integral equations, *J. Integral Equations*, **8** (1985) 113–136.

[20] S. GEORGE and M.T. NAIR, Parameter choice by discrepancy principles for ill-posed problems leading to optimal convergence rates, *J. Optim. Th. and Appl.*, **83** (1994) 217–222.

[21] S. GEORGE and M.T. NAIR, On a generalized Arcangeli's method for Tikhonov regularization with inexact data, *Numer. Funct. Anal. and Optim.*, **19** (7&8) (1998) 773–787.

[22] I.G. GRAHAM and I. SLOAN, On compactness of certain integral operators, *J. Math. Anal. and Appl.*, **68** (1979) 580–594.

[23] I.G. GRAHAM and I.H. SLOAN, Iterated Galerkin versus iterated collocation for integral equations of the second kind, *IMA J. Numer. Anal.*, **5** (1985) 355–369.

[24] C.W. GROETSCH, Comments on Morozov's discrepancy principle, In: *Improperly Posed Problems and Their Numerical Treatment*, Eds. G.Hammerlin and K.H.Hoffmann, Birkhauser, Basel, 1983, pp. 97–104.

[25] C.W. GROETSCH, *The Theory of Tikhonov Regularization for Fredholm Equations of the First Kind*, Pitman Publishing, Boston, London, Melbourne, 1984.

[26] C.W. GROETSCH, *Inverse Problems in Mathematical Sciences*, Vieweg, Braunschweg, Wiesbaden, 1993.

[27] C.W. GROETSCH, Convergence of a regularized degenerate kernel method for Fredholm integral equations of the first kind, *Integr. Equat. Oper. Th.*, **13** (1990) 67–75.

[28] C.W. GROETSCH and E. SCHOCK, Asymptotic convergence rate of Arcangeli's method for ill-posed problems, *Applicable Analysis*, **18** (1984) 175–182.

[29] J. HADAMARD, *Lectures on the Cauchy Problem in Linear Partial Differential equations* Yale University Press, New Haven, 1923.

[30] S. JOE, Collocation methods using piecewise polynomials for second kind integral equations, *J. Comp. Appl. Math.*, **12&13** (1985) 391–400.

[31] T. KATO, *Perturbation Theory for Linear Operators*, Springer–Verlag, New York, Berlin, Heidelberg, Tokyo 1976.

[32] S. KESAVAN, *Topics in Functional Analysis*, Wiley Eastern, New Delhi, 1989.

[33] J.G. KIRKWOOD and J. RISEMAN, The intrinsic viscosities and diffusion

constants of flexible macromolecules in solution, *J. Chem. Phys.*, **16** (1948) 565–573.

[34] A. KIRSCH, *An introduction to the Mathematical Theory of Inverse Problems*, Springer, New York, Berlin, Heidelberg, Tokyo 1996.

[35] R. KRESS, *Linear Integral Equations*, Springer–Verlag, New York, Berlin, Heidelberg, Tokyo 1989.

[36] J. LOCKER and P.M. PRENTER, Regularization with differential operators, *J. Math. Anal. Appl.*, **74** (1980) 504–529.

[37] S.H. KULKARNI and M.T. NAIR, A characterization of closed range operators, *Indian J. Pure and Appl. Math.* **31** (4) (2000) 353–361.

[38] B.V. LIMAYE, *Spectral Perturbation and Approximation with Numerical Experiments*, Proceedings of the Centre for Mathematical Analysis, Australian National University, Vol. 13, 1986.

[39] A.K. LOUIS, *Inverse und Schlecht Gestellte Probleme*, Teubner–Verlag, Stuttgart, 1989.

[40] E.R. LOVE, The electrostatic field of two equal circular co-axial conducting discs, *Quart. J. Mech. Appl. Math.*, **2** (1949) 428–451.

[41] C.A. MICCHELLI and T.J. RIVLIN, A survey of optimal recovery, In: *Optimal Estimation in Approximation Theory*, Eds., C.A.Micchelli and T.J.Rivlin, Plenum press, 1977, pp. 1–53.

[42] V.A. MOROZOV, On the solution of functional equations by the method of regularization, *Soviet Math. Dokldy*, **7** (1966) 414–417.

[43] M.T. NAIR, A modified projection method for equations of the second kind, *Bull. Austral. Math. Soc.*, **36** (1987) 485–492.

[44] M.T. NAIR, On strongly stable approximations, *Austral. Math. Soc. (Series A)*, **52** (1992) 251–260.

[45] M.T. NAIR, A generalization of Arcangeli's method for ill-posed problems leading to optimal rates, *Integr. Equat. Oper.*, **15** (1992) 1042–1046.

[46] M.T. NAIR, On Uniform convergence of approximation methods for operator equations of the second kind, *Numer. Funct. Anal. and Optim.*, **13** (1992) 69–73.

[47] M.T. NAIR, On accelerated refinement methods for operator equations of the second kind, *J. Indian Math. Soc.*, **59** (1993) 135–140.

[48] M.T. NAIR, A unified approach for regularized approximation methods for Fredholm integral equations of the first kind, *Numer. Funct. Anal. and Optim.*, **15** (1994) 381–389.

[49] M.T. NAIR, On Morozov's method for Tikhonov regularization as an optimal order yielding algorithm, *Zeit. Anal. und ihre Anwend.*, **18** (1999) 37–46.

[50] M.T. NAIR, Optimal order results for a class of regularization methods using unbounded operators, *Integr. Equat. Oper.*, **44** (2002) 79–92.

[51] M.T. Nair, *Functional Analysis: A first Course*, Prentice-Hall of India, New Delhi, 2002.

[52] M.T. NAIR and R.S. ANDERSSEN, Superconvergence of modified projection method for integral equations of the second kind, *J. Integral Equations*, **3** (1991) 255–269.

[53] M.T. NAIR, M. HEGLAND and R.S. ANDERSSEN, The trade–off between

regularity and stabilization in Tikhonov regularization, *Math. Comp.*, **66** (1997) 193–206.

[54] M.T. NAIR and S.V. PEREVERZEV, Regularized collocation method for Fredholm integral equations of the first kind, *J. Complexity*, **23** (2007) 454–467.

[55] M.T. NAIR, S.V. PEREVERZEV and U. TAUTENHAHN, Regularization in Hilbert scales under general smoothing conditions, *Inverse Problems*, **21** (2005) 1851–1869.

[56] M.T. NAIR and M.P. RAJAN, On improving accuracy for Arcangeli's method for solving ill-posed equations, *Integr. Equat. Oper. Th.*, **39** (2001) 496–501.

[57] M.T. NAIR and M.P. RAJAN, Arcangeli's discrepancy principle for a modified projection scheme for ill-posed problems, *Numer. Funct. Anal. and Optim.*, **22** (1&2) (2001) 177–198.

[58] M.T. NAIR and M.P. RAJAN, Arcangeli's type discrepancy principles for a class of regularization methods using a modified projection scheme, *Abstract and Appl. Anal.*, **6** (2001) 339–355.

[59] M.T. NAIR and M.P. RAJAN, Generalized Arcangeli's discrepancy principles for a class of regularization methods for solving ill-posed problems, *J. Inv. Ill-Posed Problems*, **10** (3) (2002) 281–294.

[60] M.T. NAIR and E. SCHOCK, On the Ritz method and its generalization for ill-posed equations with non-self adjoint operators, *J. International J. Pure and Appl. Math.*, **5** (2) (2003) 119–134.

[61] M.T. NAIR, E. SCHOCK and U. TAUTENHAHN, Morozov's discrepancy principle under general source conditions, *Zeit. Anal. und ihere Anwend.*, **22** (1) (2003) 199–214.

[62] A. NEUBAUER, An a posteriori parameter choice for Tikhonov regularization in the presence of modeling error, *Appl. Numer. Math.*, **14** (1988) 507–519.

[63] R. PLATO and G. VAINIKKO, On the convergence of projection methods for solving ill-posed problems, *Numer. Math.*, **57** (1990) 63–79.

[64] S. PROSSDORF and B. SILBERMANN, *Numerical Analysis for Integral Equations and Related Operator Equations*, Birkhäuser Verlag, Basel, 1991.

[65] M.P. RAJAN, *On Regularization and Approximation of Linear Ill-Posed Problems*, Ph.D Thesis, IIT Madras, March 2001.

[66] W. RUDIN, *Real and Complex Analysis*, 3rd Edition, McGraw–Hill International Editions, New York, 1098.

[67] E. SCHOCK, Uber die Konvergengeschwindigkeit projektiver verfahren I, *Math Z.*, **120** (1971) 148–156.

[68] E. SCHOCK, Uber die Konvergengeschwindigkeit projektiver verfahren II, *Math Z.*, **127** (1972) 191–198.

[69] E. SCHOCK, Galerkin–like methods for equations of he second kind, *J. Integral Equations*, 4 (1982) 361–364.

[70] E. SCHOCK, Parameter choice by discrepancy principles for the approximate solution of ill-posed problems, *Integr. Equat. Oper. Th.*, **7** (1984) 895–898.

[71] E. SCHOCK, Arbitrarily slowly convergence, uniform convergence and superconvergence of Galerkin–like methods, *IMA J. Numer. Anal.*, **5** (1985) 153–160.

[72] I.H. SLOAN, Four variants of Galerkin–like method for integral equations of the second kind, *IMA J. Numer. Anal.*, **4** (1984) 9–17.

[73] I.H. SLOAN and V. THOMEE, Superconvergence of the Galerkin iterates for for integral equations of the second kind, *J. Integral Equations and Appl.*, **9** (1985) 1–23.

[74] I.H. SLOAN, Superconvergence, In *Numerical Solution of Integral Equations*, Ed.: M.A. Golberg, Plenum Press, New York, Vol. 42 (1990) 35–70.

[75] G. VAINIKKO and P. UBA, A piecewise polynomial approximation to the solution of an integral equation with weakly singular kernel, *J. Ausutral. Math. Soc. Ser. B* , **22** (1981) 431–438.

[76] B. WENDROFF, *Theoretical Numerical Analysis*, Academic Press, New York, 1961.

[77] A.G. WESCHULZ, What is the complexity of Fredholm problem of the second kind?, *J. Integral Equations Appl.* **9** (1985) 213–241.

Index